[英国]苏布拉塔·达斯古普塔 著　姚懿容 译

牛津通识读本·

计算机科学

Computer Science

A Very Short Introduction

译林出版社

图书在版编目（CIP）数据

计算机科学 /（英）苏布拉塔·达斯古普塔
(Subrata Dasgupta) 著；姚懿容译. -- 南京：译林出
版社，2024.9. --（牛津通识读本）. -- ISBN 978-7
-5753-0286-9
Ⅰ. TP3
中国国家版本馆CIP数据核字第2024GK1892号

Computer Science: A Very Short Introduction, First Edition by Subrata Dasgupta.
Copyright © Subrata Dasgupta 2016
Computer Science: A Very Short Introduction, First Edition was originally published in English in 2016. This licensed edition is published by arrangement with Oxford University Press. Yilin Press, Ltd is solely responsible for this bilingual edition from the original work and Oxford University Press shall have no liability for any errors, omissions or inaccuracies or ambiguities in such bilingual edition or for any losses caused by reliance thereon.
Chinese and English edition copyright © 2024 by Yilin Press, Ltd
All rights reserved.

著作权合同登记号　图字：10-2020-535 号

计算机科学　[英国] 苏布拉塔·达斯古普塔 ／ 著　姚懿容 ／ 译

责任编辑　陈　锐
装帧设计　景秋萍
校　　对　施雨嘉
责任印制　董　虎

原文出版　Oxford University Press, 2016
出版发行　译林出版社
地　　址　南京市湖南路 1 号 A 楼
邮　　箱　yilin@yilin.com
网　　址　www.yilin.com
市场热线　025-86633278
排　　版　南京展望文化发展有限公司
印　　刷　江苏凤凰通达印刷有限公司
开　　本　890 毫米 ×1260 毫米　1/32
印　　张　9.5
插　　页　4
版　　次　2024 年 9 月第 1 版
印　　次　2024 年 9 月第 1 次印刷
书　　号　ISBN 978-7-5753-0286-9
定　　价　39.00 元

版权所有·侵权必究

译林版图书若有印装错误可向出版社调换。质量热线：025-83658316

序 言

王崇骏

数据与计算如同双螺旋线结构中的两条主链,围绕人类文明进程与经济社会发展这一轴心盘旋,并以时间为序不断向前发展,两者相互依存、相互影响。随着数据与计算技术的飞速发展,它们不仅仅在科学研究与实际应用中起到辅助与支撑的作用,而且可以依靠其自身的逻辑方法,驱动甚至引领科学研究和实际应用的发展。

早期,人们利用自然规律进行计时(计算),如通过圭、日晷、铜壶等计时器测得时刻;随后,人类用算盘、算筹等工具进行计算;后来,出现了利用机械装置帮助计算的计算工具。二进制(1679年)、布尔代数(1847年)和开关电路理论(1938年)的发明,为电子计算奠定了基础;1936年,图灵机的概念为电子计算提供了理论支撑;1942年,阿塔纳索夫-贝瑞计算机的发明开启了电子计算时代的探索;随后在1946年,电子数字积分计算机的发明则标志着全电子计算机时代的到来……

计算的发展历史就是一个不断演进的过程,而当代的量子计算、生物计算等新兴计算方式,也都表明人类在不断探索新的计算方式,并因此推动着科技的进步。

在这个过程中,电子计算机的发明让计算(机)本身成为研究对象,而从这时起,计算机科学就作为一个独立的科学体系得到人们的关注。所谓计算机科学,就是对计算机及其相关现象的研究。前者研究计算本身,包括计算的理论基础、方法和技术;后者关注计算在各领域中的应用,并通过多学科的交叉融合推动计算机技术的不断创新。如今,计算机科学正在成为一门前沿和引领性的学科,它不仅代表着人类智慧的结晶,也引领着社会的进步和发展,因此备受社会各界的关注和重视。

一般而言,可以从以下几个角度来理解计算机科学。

1. 计算与计算机

计算的本质在于符号处理,即以符号结构(信息、数据、知识等)作为输入,并产生符号结构作为输出。计算不仅存在于计算机系统中,也可以发生在人脑、生物系统等其他环境中。电子计算机是实现计算的工具和平台,是人工制造的产物。因此,计算机科学就是研究这种人工制品的科学。关于计算与计算机的内容详见本书的第一章和第二章。

2. 算法与编程

算法思维是解决问题的基本思维方式,启发式计算则是在算法思维基础上发展的一种特殊求解方法,用于处理复杂问题。算法描述了如何有效地完成某项任务或解决某个问题,构成了一种程序性但客观的知识。编程既是一项工程活动,也是一门数学科学,旨在将算法思维转化为可执行的代码,是算法实现的具体手段。关于算法和编程的内容,详见本书的第三章、第四章和第六章。

3. 计算机体系结构

计算机体系结构涵盖了计算机硬件和软件组织的基本结构和设计原则,描述了计算机系统的基本组成部分以及它们之间的交互方式,包括物理计算机和系统程序员之间的接口,如指令集体系结构;也包括物理(硬件)组件的实现,如处理器结构、存储系统结构、总线结构、输入/输出系统结构等。关于计算机体系结构的内容,详见本书的第五章。

4. 计算思维

问题建模与求解是计算机科学的核心问题之一,关注如何将具体问题形式化,并找到高效的解决方案。计算思维是一种通用的问题分析和解决能力:一方面,问题建模为计算思维提供具体的场景;另一方面,计算思维则指导问题建模和求解过程,两者相互依赖和互为补充。关于计算思维的内容,详见本书的第七章。

总体而言,也正如本书所述,计算机科学有一种特殊性,它吸引了人们的注意力,并将其与所有其他科学区分开来。我由衷地希望,也相信,通过阅读这本书,读者能够进一步认识和理解这个学科的魅力所在,感受到它给人类社会带来的巨大影响力。

献给安东·里彭

目 录

前　言　1

致　谢　5

第一章　计算的"特征"　1

第二章　计算人工制品　13

第三章　算法思维　32

第四章　编程的艺术、科学和工程　61

第五章　计算机体系结构的学科　78

第六章　启发式计算　100

第七章　计算思维　114

结语　计算机科学是一门通用科学吗？　122

索　引　125

英文原文　139

前　言

20世纪60年代是一个社会和文化动荡的时代。冷战、民权运动、反战示威、女权运动、学生反抗、"权力归花运动"、静坐抗议和左翼激进叛乱，但隐藏其中——几乎没有被注意到——有一门新科学在西方的大学校园里应运而生，甚至在非西方世界的一些地区虽更为初步但也出现了。

这门科学的核心是一种新型机器：电子数字计算机。围绕这台机器的技术有多种名称，最常见的是"自动运算"、"自动计算"或"信息处理"。在讲英语的国家与地区，这门科学被称为"computer science"，而在欧洲，它被称为"informatique"或"informatik"。

自动计算的技术理念——设计和制造能够在最少人工干预情况下进行计算的真实机——至少可以追溯到19世纪初英国数学家查尔斯·巴贝奇的痴迷梦想。20世纪30年代末，英国逻辑学家艾伦·图灵和美国逻辑学家阿隆佐·丘奇首先研究了计算的数学概念。但是，真正的计算经验科学的推动力要等到20世纪40年代发明、设计和实现电子数字计算机之后，也就是第二次世界大战刚刚结束之后。即便如此，此时仍存在一段酝酿期。直到20世纪60年代，大学开始提供计算机科学的本科和研究生学位，第一代受过正规培训的计算机科学家走出校园，一门拥有自己名称和身份的自主科学才开始出现。

自1946年电子数字计算机问世以来，与这台机器相关的技术（现在一般称为"信息技术"或"IT"）以及相关的文化和社会变革（一般表述为"信息时代""信息革命""信息社会"）的惊人增长是有目共睹的。事实上，我们几乎被这种技术社会环境吞没了。然而，作为以技术为基础的科学——知识学科——在专业的计算机科学社群之外是不太可见的，当然也不太为人所知。然而，计算机科学无疑与分子生物学和认知科学一样，是第二次世界大战后最重要的新科学之一。此外，计算机科学有一种特殊性，它吸引了人们的注意力，并将其与所有其他科学区分开来。

我写这本书的目的，是为那些对科学思想和原理非常感兴趣的求知欲强的读者提供理解计算机科学基本性质的基础，以丰富对这门奇怪的、历史上独一无二的、意义重大的、仍然崭新的科学的理解。简而言之，本书力求以直接、简洁的方式回答这个问题：什么是计算机科学？

在我们继续讨论之前，需要澄清一些术语。在本书中，我将使用"计算"这个词作为动词来表示某种活动；"计算结果"用作名词，表示计算的结果；"计算的"被用作形容词；"计算机"是一个名词，指的是进行计算的设备、人工制品或系统；"人工制品"指的是人类——有时是动物——制造的任何东西；"计算人工制品"是任何参与计算工作的人工制品。

最后,必须声明一个警告。本书首先接受这样一个命题:计算机科学确实是一门科学;也就是说,它体现了与科学概念相关的广泛属性,值得注意的是,它包含了对某种现象的本质进行探索的经验、概念、数学和逻辑、定量和定性模式的系统融合。对这一假设的质疑属于科学哲学的范畴,它超出了本书的范围。这里感兴趣的持久问题是计算机科学作为科学的本质,尤其是其独特而显著的特征。

致 谢

感谢牛津大学出版社的编辑拉莎·梅农,她从一开始就对这个项目给予了支持和明智的建议。她对这本书上一版的评论特别有见地。

感谢珍妮·纽吉在这项工作的各个阶段始终乐于提供编辑帮助和建议。

四位匿名读者对手稿的两份不同草稿提出了宝贵的建议和意见,这对我很有帮助。我非常感激他们,但愿我能知道他们的名字。

感谢埃尔曼·巴沙尔准备的插图。

这些材料的一部分被用在了一门关于"计算思维"的本科高年级课程中,我曾在多个场合向非计算机科学专业的学生讲授这门课程。他们的反馈对文本的塑造和精炼非常有帮助。

最后,一如既往地感谢我的家人。她们都在以不同的方式继续提供养分,使我的精神生活变得有价值。

第一章

计算的"特征"

什么是计算机科学?1967年,该学科的三位杰出早期贡献者——艾伦·佩利、艾伦·纽厄尔和赫伯特·西蒙——提供了一个现在已成为经典的答案,他们很简单地指出,计算机科学是对计算机及其相关现象的研究。

这是一个相当直接的回答,我认为大多数计算机科学家会接受它作为一个粗略且现成的工作定义。它以计算机本身为中心,当然,没有计算机就不会有计算机科学。但是,计算机科学家和好奇的外行人可能希望更精确地理解这一定义中的两个关键术语:"计算机"及其"相关现象"。

一种叫作"计算机"的自动机

计算机是一台自动机。这个词在17世纪被创造出来(复数形式为"automata"),指的是在没有外部影响的情况下,主要由自己的原动力驱动,执行某些重复的运动和动作模式的人工制品。有时,它也会模仿人类和动物的动作。早在基督以前的古代，人们就设计出了精巧的机械自动装置，主要是为了满足富人的娱乐需求，但有些具有非常实用的性质，例如，据说水钟就是公元1世纪亚历山大的工程师希罗发明的。15世纪在意大利发明的重量驱动的机械时钟，是这种自动机的一个非常成功和持久的派生物。在18世纪的工业革命中，由托马斯·纽科门（在1713年）发明，后来由詹姆斯·瓦特（在1765年）等人改进的"常压"蒸汽机，能驱动水泵从矿井中抽水，它是另一个实用的自动机实例。

因此，能够执行这样或那样物理动作的机械自动机有着悠久的历史；而模仿认知行为的自动机出现的时间则要晚得多。一个著名的例子是，20世纪40年代末至50年代初，英国神经生理学家格雷·沃尔特发明的"乌龟"机器人——"机器冒险者"。但是，在20世纪40年代后半期发展起来的自动电子数字计算机，则标志着一个全新的自动机类的诞生过程；因为计算机是用来模拟和模仿某些人类思维过程的人工制品。

将计算视为模仿人类思维的一种方式——将计算机视为"会思考的机器"——是一个非常有趣、令人不安和有争议的概念，我将在后面讨论这个概念，因为它是计算机科学的一个分支人工智能（AI）的根。但是，许多计算机科学家更愿意在他们的学科中不那么以人类为中心，有些人甚至否认计算与自主人类思维有任何相似之处。19世纪40年代，查尔斯·巴贝奇的同事、英国著名数学家阿达·洛芙莱斯伯爵夫人就指出，巴贝奇设想的机器（称之为分析机，一个世纪后成为现代通用数字计算机的第一个化身）没有"自诩"可以自行启动任务，它只能做人类命令它做的事情。现代人工智能怀疑论者经常重复这种观点，比如电子计算机的先驱之一莫里斯·威尔克斯爵士。他在20世纪末发表了一篇与洛芙莱斯的设想类似的文章，他坚持认为计算机只做"写下的程序要它们做的事情"。

那么，是什么让计算机与其他所有类型的人工制品（包括其他类型的自动机）区分开来的？是什么让计算机科学成为一门独特的科学学科的？

在本章中，我将把计算机视为"黑盒子"。也就是说，我们或多或少会忽略计算机的内部结构和工作原理，这些会在后面讨论。目前，我们将把计算机视为一种通用的自动机，只考虑它做了什么，而不考虑它是如何做这些事情的。

计算作为信息处理

每一门渴望成为"科学"的学科，都受到其所关注的基础知识的限制。物理学的内容包括物质、力、能量和运动，化学是原子和分子，遗传学的本质就是基因，土木工程则包括使物理结构保持平衡的力。

计算机科学家普遍认为，计算机科学的基本内容是信息。因此，计算机是自动从"环境"中检索、存储、处理或转换信息，并将其释放回环境的工具。这就是为什么计算的另一个术语是信息处理；为什么在欧洲，计算机科学被称为"informatique"或"informatik"；以及为什么计算界的"联合国"被称为国际信息处理联合会（IFIP）。

问题是，尽管IFIP在1960年成立（因此对信息处理的概念给予了官方的国际支持），但直到今天，关于什么是信息仍然存在很多误解。正如莫里斯·威尔克斯曾经说过的那样，这是一件难以捉摸的事情。

"无意义的"信息

造成这种情况的一个重要原因是一个不幸的事实："信息"一词被通信工程师用来表示与日常含义截然不同的东西。我们通常认为，信息会告诉我们关于世界的一些事情。在普通语言中，信息是有意义的。"X国的冬季平均气温是5摄氏度"这一陈述告诉了我们有关X国气候的一些情况，提供给我们关于X国的信息。相比之下，在通信工程学一个被称作"信息论"的分支中，信息这一词汇主要是由美国电气工程师克劳德·香农在1948年创立的，具体而言，它只是一种通过诸如电报线和电话线等通信渠道传输的商品。在信息论中，信息是没有任何意义的。信息论中的信息单位称为位或比特（二进制数字的简称），位只有两个值，通常表示为"0"和"1"。然而，在个人电脑和笔记本电脑的时代，人们对字节的概念更加熟悉。一个字节由8位组成。由于每个位可以有两个值中的一个，所以一个字节的信息可以有 2^8（= 256）个可能值，范围从00000000到11111111。不过，在"信息"这个意义上，位（或字节）的含义是无关紧要的。

在计算中，这种无意义层面的信息处理肯定是相关的，因为（后续将会介绍）由电子电路、磁性和机电设备等（统称为"硬件"）组成的物理计算机，是以位和字节的倍数来存储、处理和传递信息的。事实上，衡量计算工件的容量和性能的方法之一就是使用位和字节。例如，我可能会购买一台具有6千兆字节内部存储器和500千兆字节外部存储器（"硬盘"）的笔记本电脑（其中1千兆字节 = 10^9字节）；或者，我们可以说计算机网络以

100兆位/秒的速率传输信息（其中1兆位 = 10^6 位）。

"有意义的"（或语义）信息

但是，物理计算机（我们将在第二章中介绍）只是一种计算人工制品。无意义信息只是计算机科学家感兴趣的一种信息。另一种更重要（可以说更有趣）的信息是有意义的信息：语义信息。这样的信息与"真实世界"相联系——从这个意义上说，与这个词的日常使用相对应。例如，当我通过我的个人计算机访问互联网时，信息处理肯定是在物理或者说"无意义"的层面发生：位从某个远程计算机（"服务器"）通过网络传输到我的机器上。但我正在搜索的信息是关于某个事件的，比如说某个人的传记。我在屏幕上阅读到的文本结果对我来说是有意义的。在这个层面上，我正在与之交互的计算人工制品是一个语义信息处理系统。

当然，这些信息几乎可以是关于物理、社会或文化环境，关于过去，关于其他人公开表达的思想和想法，甚至关于一个人自己的想法（如果它们碰巧被记录或存储在某处的话）的任何东西。计算机科学家保罗·罗森布鲁姆曾经指出，这些有意义的信息与无意义的信息共享的是，这些信息必须用某些物理介质来表达，比如电信号、磁态或纸上的标记；而且它解决了不确定性。

信息是知识吗？

我们来考虑一个语义信息项，例如个人传记。在阅读它时，我就可以断言我掌握了关于那个人的知识。这指出了普通语言中信息概念的第二个混淆来源：信息与知识的混淆。

诗人T. S.艾略特对它们的区别毫不怀疑。在他的戏剧《岩石》(1934)中,他曾经提出一个著名的疑惑:

我们在知识中失去的智慧在哪里?
我们在信息中丢失的知识在哪里?

艾略特显然暗示了一种等级制度:智慧高于知识,知识高于信息。

计算机科学家通常避免谈论智慧,因为这超出了他们的职权范围。但至少在某些情况下,他们在区分知识和信息方面仍然有些不安。例如,在计算机科学的一个分支领域人工智能中,一个长期存在的问题是知识表示——如何在计算机内存中表示关于世界的知识。他们研究的另一种问题是,如何从知识体系中做出推断。人工智能研究人员认为的知识,包括事实("人皆有一死")、理论("自然选择导致的进化")、法则("每一个作用力都有一个相等和相反的反作用力")、信仰("有一个上帝")、规则("总会在停车标志处停下来")和程序("如何制作海鲜秋葵汤")等等。但是,这些实体是如何构成知识而不是信息的,这在很大程度上还没有被说明。人工智能研究人员可能会宣称,他们在计算机科学分支中所做的是知识处理而不是信息处理;但他们似乎不愿解释为什么他们关心的是知识而不是信息。

另一个被称为"数据挖掘"的专业,关注的是从大量数据中进行"知识发现"。一些数据挖掘研究人员将知识描述为隐藏在大型数据库中的"有趣"和"有用"的模式或规律。他们将知识发现与信息检索(另一种计算活动)区分开来,后者关注的是基于某些查询从数据库中检索"有用的"信息,而前者识别的知识不仅仅是"有用的"信息,或者说不仅仅是规律性的模式:这些信息必须在某种意义上是"有趣的"才能成为知识。和艾略特一样,数据挖掘研究人员认为知识优于信息。无论如何,数据挖掘的重点是知识处理,而非信息检索。

计算哲学家卢恰诺·弗洛里迪对信息和知识的关系提出了以下观点。信息和知识具有"家族相似性"。它们都是有意义的实体,但它们的不同之处在于,信息元素像砖块一样是孤立的,而知识将信息元素相互关联,这样人们就可以通过关系产生新的推论。

举个例子:假设在开车时,我在汽车收音机上听到日内瓦的物理学家检测到了一种叫作希格斯玻色子的基本粒子。这个新事实("希格斯玻色子存在")对我来说无疑是一条新信息。我甚至可能认为我获得了一些新知识。但其实这是一种错觉,除非我可以将这些信息与其他有关基本粒子和宇宙学的相关信息联系起来;而且,我也无法判断这些信息的重要性。物理学家掌握了关于亚原子粒子和宇宙结构的事实、理论、定律等的集成网络,使他们能够吸收这一新事实并掌握其意义或后果。他们拥有这样做的知识,而我只是获得了一条新信息。

信息是数据吗?

在提到数据挖掘时,我引入了另一个非常相关的术语:数据。尤其是在计算机科学界,这个术语是我们对信息概念产生模棱两可理解的一个原因。

早在1966年,计算机科学家高德纳就曾评论过,这种模棱两可的理解，本质上是一种混乱的理解。当时，计算机科学本身就是一门科学学科，它要求发明新概念并澄清旧概念。高德纳指出，在科学中，"信息"和"数据"这两个术语似乎存在一些混淆。当科学家执行一项涉及测量的实验时，引发的可能是以下四种实体中的任何一种：被测量的"真实"值；实际获得的值——真实值的近似值；值的表示；以及科学家通过分析测量得出的概念。高德纳断言，"数据"一词最适用于这些实体中的第三个。因此，对于计算机科学家高德纳而言，数据是通过以某种精确方式观察或测量所获信息的表示。因此，在他看来，信息先于数据。在实践中，信息和数据之间的关系就像信息和知识之间的关系一样模糊不清。在这里，我只能列举一些关于这种关系的不同观点。

对于著名的系统和管理科学家罗素·阿科夫来说，数据构成了观察的结果。它们是对象和事件的表示。至于信息，阿科夫想象有人问一些数据问题，然后这些数据会被"处理"（可能是由人或机器）以提供答案，而后者就是信息。因此，根据阿科夫的说法，与高德纳相反，数据是先于信息的。

对于卢恰诺·弗洛里迪来说，数据也是先于信息的，但他是在不同的意义上讨论的。根据弗洛里迪的说法，只有当系统的两种状态之间缺乏一致性时，数据才会存在。正如他所说，只要有两个变量 x 和 y，使得 $x \neq y$，就存在数据。因此，对于弗洛里迪来说，数据是一种条件，它本身没有任何意义，只是表示差异的存在。例如，当我接近红绿灯时，我观察到的红色信号是一个数据，因为它本来可以是其他情况：黄色或绿色。

鉴于数据的这一定义，弗洛里迪将信息定义为一个或多个根据某些规则结构化且有意义的数据元素。用语言学家的行话来说,当信息同时具有语法和语义时,它就是数据。因此,我对红色交通信号的观察——一个数据——变成了信息,因为红灯的含义是"驾车者必须在红绿灯处停车"。如果我不将此动作与红灯联系起来,后者将只是一个数据。

最后一个例子,对于人工智能研究人员杰弗里·施拉格和帕特·兰利来说,数据不是观察得来的。相反,观察就是数据;更准确地说,观察到的东西被选择性地记录下来以作为数据。信息是不在他们讨论之中的。

程序员的观点

这些示例足以从不同的角度说明信息和数据连接的模糊性。但是现在,让我们回到高德纳的观点。我认为,他对数据的定义在很大程度上反映了那些计算机科学家的观点,他们专门研究计算机科学的另一方面,即计算机编程——人类将计算任务传达给计算机的技术(我将在本书后面讨论这个话题)。即使在口头上支持计算作为信息处理的想法,程序员和编程理论家通常也不会考虑"信息"。相反,他们更关心像高德纳一样的科学家们的数据理念。更准确地说,他们将数据视为执行计算的基本对象("数据对象");因此,他们专注于数据对象的分类("数据类型"),表示复杂数据对象("数据结构")的规则,以及操作、处理和转换此类数据对象以产生新的数据对象的规则。对于这样的计算机科学家来说,重要的是数据,而不是信息,也不是知识。更准确地说,程序员理所当然地认为"现实世界"中存在信息。但对他们来说,更有趣的问题是,如何以一种不仅适用于自动计算而且适用于人类理解的形式来表示现实世界的信息。(毋庸赘言,像历史学家、统计学家和实验科学家这样的其他从业者,通常不会以这种方式看待数据。)

我将在本书后面详细说明这一点。但是,先举一个程序员的数据观点的非常简单的例子。在大学环境中,注册办公室存有注册学生的相关信息:他们的姓名、出生日期、家庭住址、电子邮件地址、父母或监护人姓名,他们主修的科目、修读的课程、获得的成绩、持有的奖学金、支付的费用等等。大学管理部门需要一个系统,将这些信息以系统的方式组织起来(即一个"数据库"),以便可以准确、快速地检索任何特定学生的相关信息,可以插入关于现有学生或新学生的新信息,可以有效地跟踪个别学生的进度,并且可以收集关于整个学生群体或某个子集的统计数据。被赋予创建这样一个系统任务的程序员并不关心信息本身,而是考虑到信息的性质,如何识别表示学生信息的基本数据对象,构造表示数据对象的数据结构,并构建一个数据库,以便满足大学管理部门所要求的计算任务。

符号结构作为共同命名

在本章开始时,我提出了这样一个命题:计算的基本内容是信息,计算机是处理信息的自动机,因此,计算机科学是一门研究信息处理的学科。

但我们也看到,对于一些计算机科学家(如人工智能研究人员)来说,计算的基本内容是知识而不是信息;对于其他人(例如程序员和编程理论家)来说,计算的基本内容是数据而不是信息。我们从以下计算文献中的术语示例中了解这三个实体的不同用法（其中一些已经出现在本章中，其他将在后面的章节中出现）：

数据类型、数据对象、数据结构、数据库、数据处理、数据挖掘、大数据……

信息处理、信息系统、信息科学、信息结构、信息组织、信息技术、信息存储与检索、信息论……

知识库、知识系统、知识表示、知识结构、知识论、陈述性知识、程序性知识、知识发现、知识工程、知识层次……

那么，我们能否将这三个实体——信息、数据、知识——简化为一个共同命名呢？事实上是可以的。计算机科学家保罗·罗森布鲁姆将信息等同于符号，我们可以对此进一步分析；就计算机科学而言，所有这三个实体都可以（并且通常是）由符号表示——或者更确切地说，由符号系统、符号结构来表示——也就是说，这些实体"象征"或表示其他实体。

符号需要一种表达它们的媒介，比如纸上的标记。举一个例子，"希格斯玻色子存在"这段文字是一个符号结构，其组成符号是指声音单位或音素的字母字符，加上"空白"符号；当它们串在一起时，代表了物理世界的某些东西。对于物理外行来说，这是一条信息；对于粒子物理学家来说，这成为她关于基本粒子的知识体系的一个组成部分。然而，物理学家能够理解这些信息的知识本身，就是一个更复杂的符号结构，它存储在她的大脑中和/或印刷在书籍和文章的文本中。高德纳认为数据是信息的表示，这意味着数据也是表示其他意指信息的符号结构的符号结构。即使是信息论中"无意义的"信息，比如位和字节，也可以用计算机中的物理符号表示,例如电压电平或磁态,或者在纸上用0和1的字符串表示。

　　因此,就其最基本的性质而言,计算的东西是符号结构。计算是符号处理。任何能够处理符号结构的自动机都是计算机。佩利、纽厄尔和西蒙建议的与计算机相关的"现象",最终都可以归结为符号结构及其处理。计算机科学归根结底是自动符号处理的科学,纽厄尔和西蒙都强调了这一观点。根据我们所属的计算机科学的特定"文化",我们可以选择将这些符号结构称为信息、数据或知识。

　　正是这个概念——计算最终是符号处理,计算机是符号处理自动机,计算机科学是符号处理的科学——使得计算机科学区别于其他学科。至于它的特别之处,将在后面的章节中解释。

第二章

计算人工制品

我们认为计算机是计算乃至计算机科学的核心,这是没有问题的;但仍有一些需要注意的事项。

首先,"计算机"的具体构成是什么还有待商榷。有些人倾向于将其视为他们每天使用的物理对象(笔记本电脑或工作场所的台式机);另一些人则认为是他们所使用的整个系统,包括电子邮件服务、文字处理、访问数据库等,都是"计算机";还有一些人将其与一个被称为图灵机的完全数学模型(本章稍后讨论)联系起来。

其次,接受计算机是一个符号处理自动机,以及其他与计算机相关的符号处理人工制品的观点,似乎与我们对"计算机"的直观看法略有不同。因此,我们应该更加折中地看待参与计算过程的人工制品,然后再使用计算人工制品这个术语来表述。在本章中,我们将会考虑计算人工制品的本质。

在第一章中,计算机(或多或少)表现为一个黑盒子。所说的一切都指称它是一个符号处理自动机:它接受符号结构(表示信息、数据或知识以及其他，视情况而定）作为输入，并产生（在它自己的推动下）符号结构作为输出。

当我们撬开这个黑盒子时，我们发现它更像是一组嵌套的盒子：里面有一个或多个更小的盒子；打开其中一个内盒会显示嵌套在其中的更小的盒子，并依此类推。当然，黑盒子的嵌套程度是有限的，我们迟早会到达最原始的盒子。自然世界和人造世界都体现了这种现象，被称为层级结构。许多物理、生物、社会和技术系统在结构上是分层的。自然等级（如生命系统）和人工等级（如文化或技术系统）之间的区别在于，科学家对前者是发现，而对后者则是发明。

现代计算机是一个分层组织的计算人工制品系统。因此，发明、理解和应用层级结构的规则和原则是计算机科学的一个分支学科。

自然领域和人工领域都存在层级结构是有原因的，而我们尤其要感激博学的科学家赫伯特·西蒙的这种洞察力。他说，分层组织是管理实体复杂性的一种手段。在西蒙的语言中，如果一个实体由许多以非平凡（即非显而易见）方式交互的组件组成，那么它就是复杂的。正如我们将看到的，计算机表现出这种复杂性，因此它也被组合成一个层级系统。计算机系统的设计者和实现者被要求根据层级结构的原则和规则来构建它们，计算机科学家则有责任去发明这些规则和原则。

组成层级结构

一般来说，层级系统由跨两个或更多层划分的组件组成。层级结构最常见的原则涉及层内和层间的组件关系。

图1描述了被称为"我的电脑"的物体（从物理上来说，这可能是台式电脑、笔记本电脑、平板电脑，甚至是智能手机。为了方便起见，假设它是前两者之一）。假设只将我的电脑用于三种任务：写文本（就像我现在所做的那样），发送电子邮件，以及通过互联网搜索（万维网）。因此，我认为它由三个计算工具组成，分别称为文本、邮件和网络搜索（图1中的层级1）。每个都是一个符号处理计算人工制品，是根据特定功能定义的（对于工具用户而言）。例如，文本提供了一个用户界面，允许我输入字符流，并提供命令来对齐页边距、设置行间距、分页、开始新段落、缩进、插入特殊符号、添加脚注和尾注、斜体和粗体等等。它还允许我使用命令将输入的字符流设置为文本，可以保存下来供以后使用和检索。

从我的角度来看，当我写一篇文章或一本书时，文本是我的电脑（就像此刻），就像当我发送电子邮件或搜索网络时，邮件或网络搜索则分别是我的电脑。更确切地说，对于我的电脑是什么，我有三种不同的、可供选择的想法。计算机科学家将这种虚幻的人工制品称为虚拟机，而创建、分析和理解这样的虚拟机是计算机科学中主要关注的问题之一。它们也构成了佩利、纽厄尔和西蒙所提到的围绕计算机的现象之一。

体系结构这个术语通常被计算机科学家用来表示计算工件的逻辑或功能结构。（计算机体系结构这个术语具有更专门的含义，稍后将讨论。）从我（或任何其他用户）的角度来看，计算工具文本有一个对我可见的特定架构：它有一个解释器，解释和执行命令；一个临时的或工作的存储器，保存的内容就是我正在撰写的文本；一个永久的或长期的存储器，把我选择保存的所有不

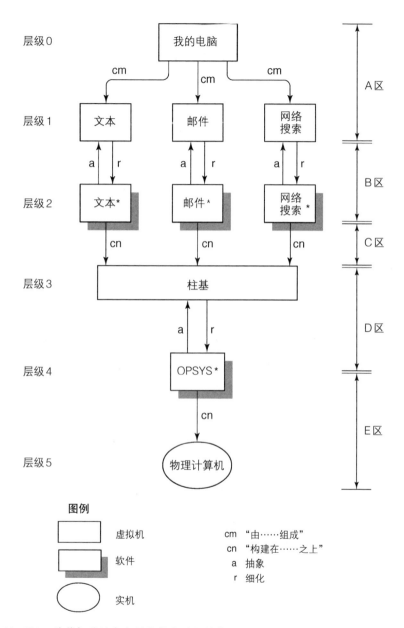

图1 计算机系统内部的抽象和层级结构

同的文本文件保存起来；输入通道将字符流和命令传输到机器，输出通道允许在屏幕上显示文本或成为印刷品（"硬拷贝"）。这些组件具有"功能性"：我可能并不知道（或特别关心）这些组件存在其中的实际媒介。因为它们描述了我（作为用户）使用文本所需要知道的一切，所以我们将其称为文本的体系结构。

同样地，当发送电子邮件时，工具邮件就是我的电脑：一个虚拟的计算人工制品，它也是一个符号处理器。邮件显示了一个用户界面，使我能够指定一个或多个消息的收件人；链接一个或多个其他符号结构（文本、图片）作为邮件的附件；撰写信息，并将其发送给收件人。邮件的体系结构类似于文本，因为它显示了相同类型的组件。它可以解释命令；有一个输入通道，使字符流能够在工作存储器中组装；有一个长期存储器，可以长期保存我想要保存的消息，以及用于在屏幕上显示电子邮件的内容并将其打印出来的输出通道。此外，邮件还可以访问其他类型的长期存储器，这些存储器保存着可以附加到消息上的符号结构（文本和图像）。但是，其中一个长期存储器是我的电脑私有的，所以只能被我访问，而另一个是公共的，也就是说，是与其他计算机用户共享的。

最后，还有网络搜索。它的架构是类似的：命令解释器；共享/公共存储（网页），其内容（网页）是可访问的；私有工作存储（临时）保存从共享存储访问的内容；私有的长期存储器用以保存这些内容，以及输入和输出通道。

图1的A部分显示的层级结构是两级的。在较高层（层级0）是单个计算人工制品，即我的电脑；但较低层（层级1）表明我的电脑是由三个独立的工具组成的。这个较低层构成了我的

工具箱。这种类型的层级结构——当一个实体A是由实体α，β，γ...组成时——在任何自然或人工的复杂系统中都是普遍存在的。这当然是计算人工制品的一个特征。由于没有一个普遍可接受的术语来描述它，所以让我们称之为组合层级结构。

抽象/细化

正如我们已经注意到的，图1的层级1中的三个计算工具在各自的架构中是相似的。每一个都包括共享和私有的长期存储器、私有的工作存储器、输入和输出通道以及命令解释器。

但是，这三个计算工具要实现，就必须使它们成为实际的工作人工制品。例如，必须有人设计并实现一个计算人工制品，当被激活时，它会作为文本执行，同时隐藏实现文本的机制细节。让我们将这个实现的人工制品表示为文本＊（图1的层级2）；它是一个计算机程序，一个软件。文本和文本＊之间是一种抽象/细化的关系（图1的B部分）：

> 一个实体E的抽象本身就是另一个实体e，它只揭示了E的那些在某些背景中被认为相关的特征，而清除了其他被认为不相关的特征（在该背景下）。相反，一个实体e的细化本身就是另一个实体E，因此E揭示了在e中缺失或被清除的特征。

文本是文本＊的抽象；反而言之，文本＊是文本的细化。注意，抽象/细化也是层级结构的一个原则，其中抽象在较高层，细化在较低层。还要注意，抽象和细化是上下文相关的。相同的实体E可以通过不同的方式进行抽象,从而产生两个或两个以上的较高层的实体e1, e2, ..., eN。反之,同样的实体e可以通过两种不同的方式进行细化,从而产生不同的较低层的实体E1, E2, ..., En。

抽象/细化原则作为管理计算人工制品复杂性的一种方式,从最早的计算年代开始就有了丰富的历史。在20世纪60年代,艾兹格·迪杰斯特拉也许是新兴的计算机科学社群最先意识到这一层级原则重要性的人。稍后我们将看到这个原则在构建计算人工制品过程中的特殊重要性,但就目前而言,读者能够了解如何用抽象/细化原则理解复杂的计算人工制品就足够了,就像可以用组合层级结构来理解复杂的计算人工制品一样。

构建的层级结构

下面我们不再从用户的角度来"看"我的电脑。我们现在所处的领域是那些真正创造了我的电脑的人:工具的制造者或设计者。顺便说一句,他们不是一个同质的群体。

特别地,文本*、邮件*和网络搜索*是计算机程序——软件,它们是程序员(现在更喜欢称呼他们为软件开发人员)在我称之为柱基(图1的C部分和第3层级)的基础架构上构建的。这个特殊的基础架构还包括文本*等的构建者可以使用的计算工具的集合。这里是第三种层级结构:构建的层级结构。

自数字计算机问世以来,设计师和研究人员就一直寻求尽可能地保护用户,使其远离真实的、有时令人讨厌的烦琐物理机器,以使用户的生活更轻松。我们的目标一直是创造一个流畅、愉快的用户界面,它接近用户的特定话语空间,并保持在用户的舒适区间。土木工程师或机械工程师想通过计算机来解决工程力学方程计算,小说家希望他的电脑可以用作书写工具,会计师想把一些比较烦琐的计算工作"转移"到计算机上,等等。在每种情况下,相关用户都希望他们的计算机是为他或她的需要量身定制的。这些年来,关于这些用户工具和基础架构是应该被合并到物理机器("硬连线")还是以更灵活的通过软件方式提供,引发了很多争论。一般来说,跨越这一鸿沟的基础架构和工具的划分,已经被计算机开发社群的特殊需求合理化了。

正如我们注意到的,我的电脑为那些只关心输入文本、收发电子邮件以及搜索网络的用户提供了这样的错觉。我的电脑为用户提供了编写文本、撰写和发送电子邮件以及在网络上搜索信息的基础架构,就像柱基(在较低层)为程序构建提供了这样的基础架构,其功能相当于用户的工具包。

但即使是通过实现文本*等程序来创建这些抽象概念的软件开发人员,也一定有自己的错觉:他们也是计算机的用户,尽管他们与计算机的接触远比我们在使用文本或邮件时更密切。我们可以称他们为"应用程序员"或"应用软件开发人员",他们也必须远离物理计算机的一些现实问题。他们也需要一个可以工作的基础架构,在这个基础架构上,他们可以创建自己的虚拟机。

在图1中,名为柱基的实体就是这样一个基础架构。事实上,它是一个程序集合(一个"软件系统")的抽象,此处显示为OPSYS*(层级4),属于一类被称为操作系统的计算人工制品。

操作系统是伟大的推动者,伟大的保护者,伟大的魔术师。在它发展的早期即20世纪60年代,它被称为"管理者"或"执行者",这些术语很好地反映了它的职责。它的功能是管理物理计算机的资源,并向计算机的所有用户(无论是外行还是软件开发人员)提供一套统一的服务。这些服务包括"加载器",它将接受要执行的程序,并将它们分配到存储器中适当的位置;存储器管理(确保一个用户程序不会侵占或干扰另一个程序使用的存储器);提供虚拟存储器(给用户一种存储器无限的错觉);控制执行输入和输出功能的物理设备(如磁盘、打印机、监视器);将储存在长期存储器中的信息(或数据或知识)进行整理,使其易于快速存取;根据标准化规则(称为"协议")执行程序,使一台计算机上的程序能够通过网络通信向另一台计算机上的程序请求服务;保护一个用户的程序不被另一个用户的程序意外或恶意损坏。图1中名为柱基的基础架构提供了这样的服务——一组计算工具;它是操作系统OPSYS*的一个抽象。

然而,操作系统并不完全是一个禁止在其上构造的程序(如邮件*)和其下的物理计算机之间所有交互的防火墙。毕竟,一个程序是通过向物理计算机发出指令或命令来执行的,而这些指令大部分被物理计算机直接解释(在这种情况下,这些指令被称为"机器指令")。操作系统所做的就是以一种受控的方式"放行"机器指令到物理计算机,并解释其他指令(例如用于输入和输出任务的指令)。

这将我们带到了(近乎)图1中所描述的层级结构的底部。操作系统软件OPSYS*,在这里显示为构建在物理计算机(层级5)之上。目前,我们将假设物理计算机(通常粗略地称之为硬件)是(最终)"真实的东西",它没有虚拟的部分。我们将看到这也是一种错觉,物理计算机有自己的内部层级,它也有自己的抽象、组合和构造层级。但至少我们可以就这一点完成目前的讨论：物理计算机提供了一个基础架构和工具箱，其中包括指令库（机器指令）、数据类型库（参见第一章）、在存储器中组织和访问指令和数据的模式，以及其他的基本设施，从而使得物理计算机能够执行程序（尤其是操作系统）。

三种类型的计算人工制品

在最近一本讲述计算机科学诞生历史的书中，我评论道，计算机科学的一个特点在于它的三类计算人工制品。

一类是材料。这些人工制品，就像历史上遇到的所有物质对象一样，遵循自然的物理定律（如欧姆定律、热力学定律、牛顿运动定律等）。它们消耗电力，产生热量，需要（在某些情况下）物理运动，随着时间的推移会产生物理和化学衰变，占用物理空间，并在运行时消耗物理时间。在图1的示例中，层级5的物理计算机是一个实例。显然，各种计算机硬件都是物质的计算人工制品。

然而，有些计算人工制品是完全抽象的。它们不仅处理符号结构，本身也是符号结构，本质上没有任何物理意义（尽管它们可以通过物理媒介，如在纸上或计算机屏幕上标记可见）。所以物理化学定律不适用于它们。它们既不占用物理空间，也不消耗物理时间。它们在物理时空中"既不劳作也不纺织"；相反，它们存在于自己的时空框架中。图1中没有抽象人工制品的实例。在下一节中，我将引用示例，并将在后面的章节中讨论其中的一些示例。但是，如果你回想起我作为一个文本或邮件用户可以设计去应用这些工具的过程，这些程序就是抽象人工制品的例证。

第三类计算人工制品是最给计算机科学带来陌生感的人工

制品。这些是抽象的和物质的。更确切地说，它们本身就是符号结构，从这个意义上说，它们是抽象的；然而，它们的运作会导致物质世界发生变化，比如信号在通信路径上传输，电磁波在空间中辐射，设备的物理状态改变，等等；此外，它们的动作依赖于底层的物质代理来执行。由于这个性质，我把这类人工制品称为阈限的（意思是一种介于两者之间模棱两可的状态）。计算机程序或软件是一大类阈限计算人工制品，例如，程序文本*、邮件*、网络搜索*和图1中的操作系统OPSYS*。

稍后，我们将遇到另一种重要的阈限人工制品。目前，使计算机科学既独特又奇怪的不仅是阈限人工制品的存在，而且我们所谓的"计算机"是材料、抽象和阈限的共生体。

在计算机科学作为一门自主的科学学科发展的大约六十年时间里，这三类计算人工制品的许多不同子类已经出现。图1显示了四个实例——用户工具和基础架构、软件和物理计算机。当然，有些子类比其他子类在计算中更重要，因为它们的作用域和使用比其他子类更普遍。此外，类和子类形成了它们自己的组合层级结构。

下面是目前在计算机科学中被认可的一些类和子类的列表。编号约定展示了它们之间的层级关系。虽然读者可能不熟悉其中的许多元素，但我将在本书中解释其中最突出的部分。

[1]抽象人工制品

 [1.1]算法

 [1.2]抽象自动机

  [1.2.1]图灵机

  [1.2.2]时序机

[1.3]元语言

[1.4]方法论

[1.5]语言

 [1.5.1]编程语言

 [1.5.2]硬件描述语言

 [1.5.3]微程序设计语言

[2]阈限人工制品

[2.1]用户工具与界面

[2.2]计算机体系结构

 [2.2.1]单处理器架构

 [2.2.2]多处理器架构

 [2.2.3]分布式计算机体系结构

[2.3]软件（程序）

 [2.3.1]冯·诺伊曼风格

 [2.3.2]功能风格

[3]材料人工制品

[3.1]物理计算机/硬件

[3.2]逻辑电路

[3.3]通信网络

"伟大的统一者"

  有一个计算人工制品必须被挑选出来描述，这就是图灵机，一个以它的创始人——逻辑学家、数学家和计算机理论家艾伦·图灵——命名的抽象机器。让我先描述一下这个人工制品，然后解释为什么它值得特别注意。

图灵机由一条被分成正方形的长度无界的纸带组成。每个方格可以容纳一个符号词汇表。在任何时间点,读写头都位于纸带的一个方格上,为"当前"方格。当前方格中的符号(包括空白符号"empty"或"black")是"当前符号"。机器可以处于有限数量状态中的一种。机器在任何给定时间的状态都是它的"当前状态"。根据当前符号和当前状态,读写头可以在当前方格上写入(输出)符号(覆盖当前符号),向左或向右移动一个方格,或实现状态改变,称为"下一个状态"。操作的循环以下一个状态作为当前状态重复,新的当前方格包含新的当前符号。(可能的)当前状态(CS),(可能的)当前(输入)符号(I),(可能的)输出符号(O),读写头(RW)的移动,以及(可能的)下一个状态(NS)之间的关系,由一个"状态表"指定。机器的行为由状态表以及影响读写、移动读写头和影响状态变化的不可见机制所控制。

图2描述了一个非常简单的图灵机,它读取写在磁带上的0

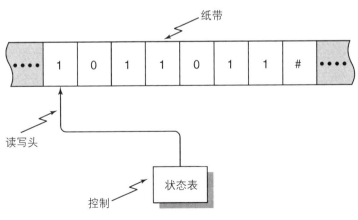

图2　图灵机的一般结构

和1的输入字符串,将输入字符串替换为0,除非整个字符串已被扫描,如果输入字符串中1的数量是奇数,则写入1,否则写为0。然后机器停止。磁带上的一个特殊符号,比如#,表示输入字符串的结尾。这台机器被称为"奇偶校验检测器":它将整个输入字符串替换为0,并将#替换为1或0,具体取决于输入字符串的奇偶校验(即1的数量)是奇数还是偶数。

这台机器需要三个状态:S_o表示在机器运行的任意时刻,在输入字符串中检测到奇数个1。S_e表示在机器运行的任意点检测到偶数个1。第三种状态H是停机状态:它使机器停止。当机器开始运行时,它的读写头指向保存输入字符串中第一个数字的方格。

图灵机的潜在行为由状态表指定(参见表1)。

表1 状态表

当前状态	输入符号	下一个状态	输出符号	移动读写头
S_e	0	S_e	0	R
S_e	1	S_o	0	R
S_e	#	H	0	—
S_o	0	S_o	0	R
S_o	1	S_e	0	R
S_o	#	H	1	—

该表中的每一行都指定了机器上的不同操作,并且必须独立解释。例如,第一行表示:如果当前状态是S_e,当前输入符号是0,那么下一个状态(也)将是S_e,输出符号为0并写入磁带,读写头向右移动一个位置。最后一行告诉我们,如果当前状态

是 S_o，输入符号是 #，那么将 # 替换为 1，并使下一个状态为暂停状态 H。读写头不发生移动。假设输入字符串如图 2 所示，机器被设置为状态 S_e。读者可以很容易地验证，在机器运行的连续周期中，磁带的状态顺序和内容如下。每个循环中读写头的位置由"当前"输入符号右侧的上标星号表示：

S_e：1*011011# → S_o：00*11011# → S_o：001*1011# → S_e：0001*011# → S_o：00000*11# → S_o：000001*1# → S_e：0000001*# → S_o：0000000#* → H：00000001*

那么，对于每一个不同的符号处理任务，就会有一个不同的图灵机（图灵自己简单地称之为"计算机"）。每个这样的（特殊用途）图灵机会指定机器所要识别的符号字母表、可能的状态集、读写头所在的初始方格、状态表和初始当前状态。在机器操作结束时（当它到达"停止"状态），写到磁带上的输出会给出符号处理任务的结果。

因此，例如，图灵机可以用来将两个数字 n 和 m 相加，n 和 m 表示为 n 个 1，接着是一个空格，然后是 m 个 1，在纸带上留下结果 $n+m$（$n+m$ 个 1 的字符串）。另一个图灵机使用由 a、b 和 c 符号组成的单个字符串作为输入，它将用输入字符串的"镜像"（称为"回文"）替换这个输入字符串。例如，如果输入字符串是 'aaabbccc'，那么输出将是 'cccbbbaaa'。因此，图灵机是一种符号处理机。当然，它是"最纯粹"意义上的抽象人工制品，因为机器本身是一个符号结构。人们很难会想到把这种物理版本的图灵机作为一种实用的人工制品。

但是，图灵想得要更长远。他还表明，人们可以建立一个单独的计算机器U，它可以模拟其他所有图灵机。如果向U提供包含特定图灵机状态表描述的磁带，U将解释该描述，并执行与该特定机器相同的任务。这样的机器U被称为通用图灵机。

图灵的发明，意义在于他提出的一个主张，即**任何我们"直观上"或"自然地"认为是计算过程的过程都可以用图灵机来实现**。由此可见，通用图灵机可以执行任何我们认为是计算的任务。这种说法被称为图灵论题，有时也被称为丘奇-图灵论题，因为另一位逻辑学家阿隆佐·丘奇用完全不同的思路得出了同样的结论。

我们可以把图灵机视为"伟大的统一者"。它绑定了所有的计算人工制品；也就是说，所有的计算人工制品及其行为都可以简化为图灵机的工作方式。

在计算机科学中存在一个被称为自动机理论的完整分支，它研究图灵机的结构和行为，研究图灵机在所有可能的表现形式中的能力和局限性（例如，把纸带限制在一个有限长度，或引入多个具有多个读写头的纸带），我们必须认识到一个矛盾的情况，即图灵机几乎没有影响到任何实际（或可行）计算人工制品的发明、设计、实现和行为，或者处理此类人工制品的计算机科学家的思维和实践！

交互式计算

此外，自图灵时代以来，已经出现了可以与其他自然或人工系统交互的计算人工制品。"交互"在这里指的是人工（包括社会）和/或自然智能体之间的彼此或相互影响，共同形成某种系统。

例如，考虑一下我支付水电费账单。它需要我和我的笔记本电脑、我的银行计算机系统以及公用事业公司的计算机系统之间的交互。在这种情况下，四个智能体（三个计算人工制品和我自己）通过交换消息、命令和数据交互式地影响信息传输和计算。

或者考虑一下图1中的抽象计算人工制品文本。这构成了一个人机界面，文本的人类用户和软件系统文本*互动。由文本提供并由用户发出的命令会导致文本*做出响应（启动新的文本行，在单词之间创建空格，在文本中添加字符以形成单词，缩进新的段落，斜体化单词，等等），而后者的响应反过来会提示人类用户做出响应。

这种交互式系统不符合图灵机的"标准"理念，图灵机本质上是一个独立的人工制品，在机器激活之前，输入已经写在纸带上，并且只有当图灵机停止时，才可以看到输出。交互式计算人工制品（例如我的银行或我的公用事业公司的系统）可能永远不会停止。

正是出于这样的考虑，一些计算机科学家坚持认为图灵机的研究——自动机理论——应该属于数学和数理逻辑的范畴，而不是严格意义的计算机科学，而另一些人则质疑图灵机理论包含整个计算的观点的正确性。

计算机科学作为一门人工科学

总结一下到目前为止的讨论，计算人工制品是人造的东西；他们处理表示信息、数据或知识的符号结构（取决于个人的观点和背景)。计算机科学是计算人工制品的科学。

很明显,计算人工制品不像岩石、矿物和化石、植物和动物、恒星、星系和黑洞、基本粒子、原子和分子那样是自然世界的一部分。人类创造了这些人工制品。因此,计算机科学不是一门自然科学。那么它是一门什么样的科学呢?

一种观点认为,由于计算人工制品是功利的,因此它是技术性的,计算机科学根本不是一门"真正的"科学。更确切地说,它是工程学的一个分支。然而,材料强度、结构理论、热力学、物理冶金、电路理论等传统工程科学,以及生物工程、基因工程等新兴工程科学,都受到自然规律的直接制约。阈限和抽象的计算人工制品似乎与工程科学家研究的不折不扣的材料人工制品——结构、机床、发动机、集成电路、金属、合金和复合材料等——相去甚远。这就是为什么材料的计算人工制品(计算机硬件)通常属于工程学院的领域,而阈限和抽象的人工制品则属于科学学院的领域。

然而,所有的人工制品,包括工程层面上的和计算层面上的,都有一些共同之处:它们是人类思想、人类目标、人类需求和欲望的产物。**人工制品具有目的性:它们反映了创造者的目标。**

赫伯特·西蒙把所有与人工制品(抽象的、阈限的或材料的)有关的科学统称为人工科学。它们与自然科学区别开来,是因为它们必须考虑目标和目的。一个自然物体是没有目的的,岩石和矿物、恒星和星系、原子和分子、植物和生物并不是带着目的来到这个世界的。他们只是存在着。天文学家不会问:"星系是干什么用的?"地质学家不会问:"火成岩侵入的目的是什么?"自然科学家的任务是发现支配自然现象结构和行为的规律,探究它们是如何产生的,但不是问它们为什么或是出于什么目的产生的。

相比之下,人工制品进入世界反映了人类的需求和目标。如果我们忽略了它们存在的原因,而仅仅探究支配计算人工制品(或者金字塔、吊桥、粒子加速器和厨房刀具)的结构和行为的法则和原则是不够的。

人工科学需要研究手段和目的之间的关系,即一件人工制品被预定的目的或需求,以及人工制品的制造是为满足需求。因此,计算机科学中的"科学"是一门关于手段和目的的科学。它提出了这样的问题:给定人类的需求、目标或目的,计算人工制品如何能够明确地实现这样的目的?也就是说,人们如何通过推理、观察或实验证明计算人工制品满足这一目的?

第三章
算法思维

就像莫里哀的戏剧中那个不知道自己一生都在讲散文的角色一样,大多数人可能没有意识到,当他们还是孩子的时候,他们第一次将两个多位数字相乘或相除时,他们正在执行一个算法。事实上,在20世纪60年代之前,除了计算和数学界,很少有人知道"算法"这个词。然而,从那时起,"算法"就像"范式"(一种最初在科学哲学中流行的术语)一样,已经进入了通用语言,用以表示解决问题的公式、规则、配方或系统程序。这在很大程度上是由于在过去五十年左右的时间里,计算与算法有着密切的联系。

然而,算法的概念(如果不是这个词的话)可以追溯到古代。欧几里得的伟大著作《几何原本》(公元前300年)阐述了平面几何的原理,描述了寻找两个正整数的最大公约数(GCD)的算法。"算法"一词起源于9世纪阿拉伯数学家和天文学家阿布·阿卜杜拉·穆罕默德·伊本·穆萨·花拉子密的名字,他生活和工作在巴格达的"智慧之家"——那个时代世界上最重要的科学中心之一。在他众多关于数学和天文学的论文中，花拉子密写到了"印度的计算艺术"。后来的拉丁译本读者误以为这是花拉子密自己的作品，称他的作品为"algorismi"，最后变成了"algorism"，意思是循序渐进的过程。这就演变成了"algorithm"（算法）。《牛津英语词典》能找到的关于这个词的最早参考文献是1695年发表在英语科学期刊上的一篇文章。

高德纳可能比其他任何人都更能让算法成为计算机科学家意识的一部分，他曾经将计算机科学描述为对算法的研究。并不是所有的计算机科学家都同意这种"总体化"的观点，但是一门没有以算法为中心的计算机科学是难以想象的。就像达尔文的生物学进化论一样，计算机领域的条条大路似乎都通向算法。如果说生物学思维是进化思维，那么计算思维就是形成算法思维的习惯。

石蕊测试

作为进入这一领域的一个入口，让我们考虑一下石蕊测试，这是学生在高中化学课中进行的首批实验之一。

试管或烧杯里有一种未知种类的液体。实验人员将一条蓝色石蕊试纸浸入其中。如果它变成红色，则液体是酸性的；若它依然是蓝色的，则说明液体不是酸性的。在后一种情况下，实验者将一条红色石蕊试纸浸入液体中，它变成蓝色，因此液体是碱性的，否则它是中性的。

这是学生在化学教育中很早就学会的一个决策过程，我们可以用以下方式描述它：

>   **if** 蓝色石蕊试纸浸入液体会变成红色
>   **then** 得出液体是酸性的结论
>   **else**
>   **if** 红色石蕊试纸浸入液体后变成蓝色
>   **then** 就可以断定该液体是碱性的
>   **else** 就可以断定该液体是中性的

其中使用的符号将在本章和后面的一些章节中出现,这里解释一下。一般来说,**if** C **then** $S1$ **else** $S2$ 的表示,在算法思维中用于指定某种类型的决策。如果条件 C 为真,那么算法中的控制流将进入 $S1$,然后 $S1$ 将执行;如果 C 为假,则控制转到 $S2$,然后 $S2$ 将执行。在这两种情况下,在执行 **if then else** 语句后,控制权都会转到算法中跟在它后面的语句。

请注意,实验者不需要知道石蕊测试为什么会这样工作。她不需要知道"石蕊"的化学成分到底是什么,也不需要知道发生了什么化学过程导致了颜色的变化。满足以下两个条件就完全足够执行这个程序:第一,实验者看到石蕊试纸时认出它;第二,她能将颜色的变化与酸和碱联系起来。

"石蕊测试"一词已经成为一种确定的条件或测试的隐喻,并且具有充分的理由:它保证有效。这代表会有一个明确的结果,没有不确定的余地。此外,石蕊测试不能无限期地进行下去,实验者确信在有限的时间内,测试会给出一个决定。

石蕊测试的这些结合特性——保证在有限时间内产生正确结果的机械程序——是表征算法的基本要素。

什么时候程序是算法？

对于计算机科学家来说，算法不仅仅是一个机械的程序或配方。为了使一个程序符合算法，就如计算机科学家对这个概念的理解，它必须具有以下属性（正如高德纳首次阐明的那样）：

（1）有限性。算法总是在有限的步骤之后终止（即停止）。

（2）确定性。算法的每一步都必须精确而明确地指定。

（3）有效性。作为算法的一部分执行的每个操作必须足够原始，以便人来精确地执行（比如使用铅笔和纸）。

（4）输入和输出。一个算法必须有一个或多个输入和一个或多个输出。

让我们考虑一下欧几里得之前提到的古老算法，求两个正整数 m 和 n 的最大公约数（也就是能被 m 和 n 整除的最大正整数）。该算法在这里用一种结合了通用英文、基本数学符号和一些用于表示决策的符号（如石蕊测试的例子）来描述这个算法。在算法中，m 和 n 分别作为"输入变量"和"输出变量"。此外，还需要第三个"临时变量"，记为 r。不属于算法本身的"注释"被包含在"{ }"中。算法中的"←"符号有特殊意义：它表示"赋值操作"。"$b \leftarrow a$"表示将变量 a 的值复制或赋值给 b。

欧几里得最大公约数算法：给定两个正整数 m 和 n 找到它们的最大公约数。

Input m, n $\{m, n \geq 1\}$;

Temp var r;

Step 1: divide m by n; $r \leftarrow$ remainder; $\{0 \leq r \leq \}$;

Step 2: if $r = 0$ **then Output** n ;

　　　　　　　　Halt

　　　　　else

Step 3: $m \leftarrow n$;

　　　　$n \leftarrow r$;

　　　goto step 1.

假设 $m = 16$, $n = 12$。如果一个人用铅笔和纸来"执行"这个算法,那么每一步执行后三个变量 m, n, r 的值将如下所示:

	m	n	r
Step1:	16	12	4;
Step3:	12	4	4;
Step1:	12	4	0;
Step2:	Output n = 4.		

另一个例子,假设 $m = 17$, $n = 14$。每个步骤执行后,三个变量的值如下:

	m	n	r
Step1:	17	14	3;
Stcp3:	14	3	3;
Step1:	14	3	2;
Step3:	3	2	2;
Step1:	3	2	1;

Step3:	2	1	1;
Step1:	2	1	0;
Step2:	Output $n = 1$.		

在第一个例子中，GCD（16，12）= 4，这是算法停止时的输出；在第二个例子中，GCD（17，14）= 1，算法在终止后输出该值。

显然，算法是有输入的。不太明显的是，算法是否满足有限性准则。有一个由 **goto** 命令指示的重复或迭代，它使控制返回到步骤1。正如这两个例子所示，该算法在步骤1和步骤3之间迭代，直到满足条件 $r = 0$，然后输出 n 的值作为结果，算法停止。这两个例子清楚地表明，这些特定的 m 和 n 输入值对，算法总是最终满足终止条件（$r = 0$），并将停止。然而，我们如何知道对于其他的值对，它是否不会永远迭代步骤1和步骤3，并且永远不会产生输出呢？（如果是这种情况，算法将同时违反有限性和输出准则。）我们怎么知道对于 m 和 n 的所有可能的正值，算法都会终止呢？

答案是必须证明，在普遍情况下，算法是有限的。这个证明是在步骤2中对 $r = 0$ 的每次检验后，r 的值都小于正整数 n，并且随着步骤1的每次执行，n 和 r 的值都在减小。一个递减的正整数序列最终必须达到0，因此最终 $r = 0$，因此通过步骤2，程序最终将终止。

下面考虑确定性标准，这意味着算法的每一步都必须被精确定义，必须明确地指定要执行的操作。因此，语言进入了画面。对欧几里得算法的描述混合了英语和模糊的数学符号。执行这个算法的人（借助铅笔和纸）应该确切地理解除法的含义，余数是什么，正整数是什么。他必须理解更正式的符号的含义，比如符号 'if ... then ... else'，'goto'。

至于有效性，所要执行的所有操作必须足够原始，以便可以在有限的时间内完成。在这种特殊情况下，指定的操作足够基本，可以像前面那样在纸上执行它们。

进一步思考与前行

抽象的概念适用于算法的规范。换句话说，一个特定的问题可以通过在两个或更多不同抽象级别上指定的算法来解决。

在袖珍计算器出现之前，孩子们被教用铅笔和纸做乘法运算。以下是我小时候学到的东西。为简单起见，假设一个三位数（"被乘数"）乘以一个两位数（"乘数"）。

步骤1：将这些数字的被乘数放在最上面一行，乘数放在最下面一行，并将乘数的个位与被乘数的个位对齐。

步骤2：在乘数下面画一条水平线。

步骤3：将被乘数乘以乘数的个位，并将结果（"部分乘积"）写在水平线的下方，使所有的个位数对齐。

步骤4：在步骤3得到的部分乘积的个位下面放置一个"0"。

步骤5：将被乘数与乘数的十位相乘，并将结果（部分乘积）放置在横线以下第二行，即"0"的左边。

步骤6：在第二个部分乘积下面再画一条水平线。

步骤7：把两个部分乘积相加，写在第二条水平线的下方。

步骤8：停止。第二行下面的数字是所需的结果。

请注意，要成功地完成这个过程，孩子必须有一些先验知识：(1) 她必须知道如何将一个多位数的数字乘以一个一位数的数字。这需要记住两个一位数乘法的乘法表，或者参照乘法表。(2) 她必须知道如何将两个或多个的一位数相加，必须知道如何进位。(3) 必须知道如何把两个多位数的数字相加。

然而，孩子不需要知道或理解为什么这两个数字按照步骤1对齐；或者，为什么第二个部分乘积按照步骤5向左移动一个位置；或者，为什么在步骤4中插入一个"0"；或者，当她在步骤7中将两个部分乘积相加时，为什么会得到正确的结果。

但是，请注意这些步骤的精确性。在严格按照规定的步骤进行的情况下，只要执行程序的人员满足前面提到的条件(1)和(2)，就可以保证该程序正常工作。它保证在有限的时间内产生正确的结果，这也是算法的基本特征。

想想我们现在大多数人会如何做这个乘法。我们将召唤我们的袖珍计算器（或智能手机），我们将继续使用计算器，步骤如下：

步骤1：输入被乘数。

步骤2：按"x"。

步骤3：输入乘数。

步骤4：按"="。

步骤5：停止。结果就是显示的内容。

这也是一个乘法算法。这两种算法实现了相同的结果，但它们处于两个不同的抽象层次。在第二种算法中，用户不知道执行步骤1到步骤4时到底发生了什么。计算器很有可能执行了与纸笔版本相同的算法，也有可能使用了不同的执行。这个信息对用户是隐藏的。

这个例子中的抽象级别也意味着无知的级别。使用纸笔算法的孩子比使用袖珍计算器的人更了解乘法。

算法的确定性

算法有一个令人欣慰的特性，其性能不依赖于执行者，只要执行者满足前面提到的知识条件（1）—（3）。对于算法的相同输入，无论什么人（或物）执行算法，都将获得相同的输出。此外，无论何时执行，算法都会产生相同的结果。总的来说，这两个属性意味着算法是确定的。

这也说明了为什么食谱通常不是算法，它们通常包含模棱两可的步骤，从而破坏了确定性标准。例如，它们可能包括添加"轻碎"或"细磨"食材的说明，"慢煮"的指令，等等。这些指令太模糊，无法满足算法确定性的条件。更确切地说，需要厨师的直觉、经验和判断来解释这些指示。这就是为什么由两个不同的厨师用同一个食谱烹制的同一道菜可能味道不同；或者，为什么同一个人在两个不同的场合遵照相同的食谱可能会做出不同的味道。食谱违反了确定性原则。

算法是抽象的产物

算法无疑是一种人工制品，它是人类根据目标或需求而设计或发明的。并且，就它们处理符号结构（就像在最大公约数和乘法算法的情况下）而言，它们是计算性的。（并非所有算法都处理符号结构。石蕊测试以物理实体——液体试管、石蕊试纸条——作为输入，并产生一种物理状态——石蕊试纸条的颜色——作为输出。石蕊测试是一种手动算法，它作用于物理化学实体,而不是符号结构;因此,我们不应该将其视为计算人工制品。)

对于算法而言,无论是不是计算性的,它们本身是没有物理存在的。人们既不能触摸它们,也不能握住、感觉、品尝它们,也听不到它们。它们既不遵守物理和化学定律,也不遵守工程学定律。它们是抽象的人工制品。它们是由符号结构组成的,与所有符号结构一样,这些符号结构代表世界上的其他事物,这些事物本身可能是物理的(石蕊试纸、化学药品、袖珍计算器上的按钮等),也可能是抽象的(整数、整数的运算、等式之类的关系)。

算法是一个工具。就像大多数工具一样的情况,用户需要知道的理论基础越少,算法对用户越有效。

这就提出了以下观点:作为人工制品,算法是双面的。它的设计或发明通常需要创造力,但它的使用是一种纯粹的机械行为,几乎不需要创造性思维。可以说,执行算法是一种无意识的思考形式。

算法是程序性知识

作为人工制品,算法是用户用来解决问题的工具。一旦创造出来并公之于众,它们就属于全世界。这就是使算法成为客观的(以及确定的)人工制品的原因。但算法也是知识的体现。作为客观的人工制品,它们是科学哲学家卡尔·波普尔所称的"客观知识"的体现。(这里我们说算法的用户既是一个"无意识的思考者",又是一个"认知主体",这听起来可能有点矛盾;但即使是盲目的思考也仍然是思考,而思考需要汲取某种知识。)

但是算法代表了什么样的知识呢？

在自然科学中，我们学习定义、事实、理论、规律等。这里有一些来自基础物理和化学的例子。

1. 真空中的光速为每秒18.6万英里。
2. 加速度是速度的变化率。
3. 氢的相对原子质量为1。
4. 物质有四种状态：固态、液态、气态和等离子态。
5. 当化学元素按原子序数排列时，元素的性质具有周期性（循环）规律。
6. 燃烧需要氧气的存在。

在每一种情况下，都有这样的说法：燃烧需要氧气的存在，氢的相对原子质量是1，加速度是速度的变化率，等等。这些陈述的（近似）真理要么通过定义（2）、计算（1）、实验或观察（4,6），要么推理（5）。严格地说，孤立地看，它们是信息项，当被吸收时，就会成为一个人的知识的一部分（见第一章）。这种知识被称为陈述性知识，或者更通俗地说，"知道"的知识。

数学也有陈述性知识，以定义、公理或定理的形式存在。例如，一个基本的算术公理，源于意大利数学家朱塞佩·皮亚诺的数学归纳法原理：

> 任何属于零的属性，以及属于具有该属性的每个数字的直接后继者，都属于所有的数字。

相比之下，毕达哥拉斯定理是平面几何中一个通过推理（证明）的陈述性知识：

在直角三角形中，a，b组成直角，c是斜边，$c^2 = a^2 + b^2$的关系是成立的。

下面是一个定义的陈述性数学知识的例子：

非负整数 n 的阶乘为：
$$factorial(n) = 1 \text{ for } n = 0 \text{ or } n = 1$$
$$= n(n-1)(n-2)...3.2.1 \text{ for } n > 1$$

相比之下，算法不是陈述性的；相反，它构成了一个过程，描述如何做某事。它规定了某种行为。因此，算法是程序性知识的一个实例，或者通俗地说，是"诀窍"。

对于一个计算机科学家来说，仅仅知道一个数的阶乘被如何定义是不够的。她想知道如何计算一个数的阶乘。换句话说，她想要一个算法。例如：

阶乘

  Input: $n \geqslant 0$;

  Temp variable: *fact*;

 Step 1: *fact* ← 1;

 Step 2: if $n \neq 0$ **and** $n \neq 1$ **then**
   repeat

 Step 3: *fact* ← *fact* * *n*;

 Step 4: $n = n - 1$;

 Step 5: until $n = 1$;

Step 6: output *fact* ;
Step 7: halt

符号 **repeat** *S* **until** *C* 指定迭代或循环。语句 *S* 将迭代执行，直到条件 *C* 为真。当这种情况发生时，循环终止，并且控制流到迭代之后的语句。

这里，在步骤1中，*fact* 被赋值为1。如果 $n = 0$ 或 1，则不满足步骤2的条件，此时控制直接转到步骤6，输出 *fact* = 1 的值，算法在步骤7停止。另一方面，如果 *n* 既不是1也不是0，那么由 **repeat ... until** 段所指示的循环将迭代执行，每次 *n* 递减1，直到满足循环终止条件 $n = 1$。然后，控制进入步骤6，执行时输出 *fact* 的值 $n(n-1)(n-2)...3.2.1$。

请注意，同样的概念——"阶乘"——既可以陈述的方式（正如数学家所喜欢的）也可以程序的方式（如计算机科学家所希望的）来表示。事实上，陈述的形式为程序的形式也就是算法（"我们如何计算它？"）提供了基础的"理论"（"什么是阶乘？"）。

总之，算法构成了一种程序性但客观的知识。

设计算法

当我们考虑算法的设计时，算法的抽象性会产生一个奇怪的结果。这是因为，一般来说，设计是一种以目标为导向（有目的性）的行为，它以一组需求 R 开始，由一个尚未存在的人工制品 A 来满足，并以表示所需人工制品的符号结构结束。通常情况下，这个符号结构就是人工制品 A 的设计 D（A），设计师的目标是创造D（A），使得如果根据D（A）实现A，那么A就会满足R。

当人工制品A是材料的时候，这种情况是没有问题的；例如，桥梁的设计会以工程图纸的形式表示桥梁的结构，以及显示作用在结构上的力的一组计算和图表。然而，在算法作为人工制品的情况下，人工制品本身是一个符号结构。因此，讨论算法的设计，就是讨论代表另一个符号结构（算法）的符号结构（设计）。这是有些令人费解的。

因此，在算法的情况下，更明智和理性的是认为设计和人工制品是一样的。设计算法的任务就是创建一个算法A的符号结构，使A满足R的要求。

这里的关键词是"创造"。设计是一种创造性行为，正如创造性研究人员所表明的那样，创造性行为是理性、逻辑、直觉、知识、判断、诡计和机缘巧合的复杂混合体。然而，设计理论家确实在谈论"设计科学"或"设计逻辑"。

那么，算法的设计是否有科学的成分呢？答案是，"在一定程度上"。从本质上讲，有三种方法可以让"科学方法"进入算法的设计。

首先，设计问题并不存在于真空中。它是由设计师所拥有的与问题相关的知识体系（称为"知识空间"）进行情景化的。在设计一个新的算法时，这个知识空间变得很重要。例如，可能会发现手头的问题与现有算法（属于知识空间的一部分）解决的问题之间的相似性；因此，用于后者的技术可以用于当前的问题。这是一个类比推理的例子。或者，一个已知的设计策略似乎特别适合手头的问题，所以可以尝试这个策略，尽管不能保证它的成功。这是启发式推理的一个例子。或者,可能存在一个与问题所属领域相关的形式理论,因此,这一理论可以用于解决这个问题。这是一个理论推理的例子。

换句话说,在基于一套已建立的或经过充分试验的或已证明的知识(陈述性的和程序性的)来设计算法时,可以采用推理的形式。我们称之为算法设计中的"知识因素"。

但是,仅仅提出一个算法是不够的,还有义务让自己和他人相信这个算法是有效的。这需要通过系统推理证明算法满足原始需求。这里我将其称为算法设计中的"有效性因素"。

最后,即使证明该算法是有效的,这也可能是不够的。有一个关于它的性能的问题:这个算法有多好?我们将其称为算法设计中的"性能因素"。

这三个"因素"都包含了我们通常与科学相关的各种推理、逻辑和证据规则。让我们通过一些例子来看看它们是如何对算法设计科学做出贡献的。

翻译算术表达式的问题

有一类计算机程序叫作编译器,它们的工作是将"高级"编程语言(即一种从实际物理计算机的特性中抽象出来的语言,例如Fortran或C++)编写的程序翻译成可以由特定物理计算机直接执行(解释)的指令序列。这种特定于机器的指令序列被称为"机器代码"。(编程语言将在第四章中讨论。)

最早的编译器作者(在20世纪50年代末和60年代)面临的一个经典问题是,开发算法,以便将程序中出现的算术表达式转换为机器代码。这种表达式的一个例子如下:

$$(a+b)*(c-1/d)$$

其中 +、−、* 和 / 是四个算术运算符；变量 a,b,c,d 和常数 1 被称为"操作数"。在这种形式的表达式中，算术运算符出现在两个操作数之间，被称为"中缀表达式"。

围绕这个问题的知识空间（算法设计者所拥有的知识空间）包括以下算术运算符的优先级规则：

1. 在没有括号的情况下，*、/ 优先于 +、−。

2. *、/ 具有相同的优先级；+、− 的优先级相同。

3. 如果表达式中出现相同优先级的运算符，则适用从左到右的优先级。也就是说，执行者按从左到右出现的顺序应用于操作数。

4. 括号内的表达式优先级最高。

因此，例如在前面给出的表达式中，运算符的顺序为：

（1）执行 $a+b$，得到结果 $t1$

（2）执行 $1/d$，得到结果 $t2$

（3）执行 $c-t2$，得到结果 $t3$

（4）执行 $t1*t3$

另一方面，如果表达式没有括号：

$$a+b*c-1/d$$

那么运算符的顺序为：

（1）执行 $b*c$，得到结果 $t1'$

（2）执行 $1/d$，得到结果 $t2'$

（3）执行 $a+t1'$，得到结果 $t3'$

（4）执行 $t3'-t2'$

可以设计一种算法来生成机器代码,该机器代码在执行时将根据优先规则正确地计算中缀算术表达式。(算法的精确性质将取决于机器相关指令的性质,这是特定物理计算机的特性。)因此,该算法基于优先规则,提取与问题相关的知识空间的精确规则。此外,由于该算法直接基于优先规则,因此将大大方便对算法的有效性论证。然而,正如前面的例子所示,括号使中缀表达式的翻译复杂了很多。

波兰逻辑学家扬·卢卡西维茨(1878—1956)发明了一种不需要括号就能指定算术表达式的表示法,因此被称为"波兰式表示法"。在这种表示法的一种形式——被称为"逆波兰式"——中,运算符紧跟在两个操作数后面,以逆波兰式表达式表示。下面的例子展示了一些中缀表达式的逆波兰式表达式的形式。

(1) $a+b$ 对应的逆波兰式是 $a\,b+$

(2) $a+b-c$ 对应的逆波兰式为 $a\,b+c-$

(3) $a+b*c$ 的逆波兰式是 $a\,b\,c*+$

(4) $(a+b)*c$ 对应的逆波兰式为 $a\,b+c*$

逆波兰式表达式的求值以一种直接的方式从左到右进行,从而使翻译问题更容易。规则是,所遇到的算术操作符按照操作符出现的顺序(从左到右)应用于其前面的操作数。例如,中缀表达式如下:

$$(a+b)*(c-1/d)$$

逆波兰式表达式的形式是

$$ab+c1d/-*$$

执行顺序为：

（1）执行 $a\,b\,+$，得到结果 $t1$。结果表达式是 $t1\,c\,1\,d\,/\,-\,*$

（2）执行 $1\,d\,/$，得到结果 $t2$。结果表达式是 $t1\,c\,t2\,-\,*$

（3）执行 $c\,t2\,-$，得到结果 $t3$。结果表达式是 $t1\,t3\,*$

（4）执行 $t1\,t3\,*$

当然，程序员会用熟悉的中缀形式编写算术表达式。编译器将实现一个算法，该算法首先将中缀表达式转换为逆波兰式表达式，然后从逆波兰式表达式生成机器代码。

将中缀表达式转换为逆波兰式表达式形式的问题说明了，如何将一个可靠的理论基础和一个经过验证的设计策略结合起来设计一个可被证明是正确的算法。

这种设计策略被称为递归，是更广泛的"分而治之"问题解决策略的一个特例。在后者中，给定一个问题 P，如果它可以划分为更小的子问题 $p1, p2, …, pn$，独立求解 $p1, p2, …, pn$，然后将子问题的解组合起来，得到 P 的解。

在递归中，问题 P 被分成许多与 P 类型相同但更小的子问题。每个子问题又被分成更小的相同类型的子问题，依此类推，直到子问题变得足够小而简单，可以直接解决。然后，子问题的解被组合起来给出"父级"子问题的解，这些子问题组合起来形成其父级问题的解，直到得到原始问题 P 的解。

现在考虑将算法中缀表达式转换为逆波兰式表达式的问题。它的基础是一套正式规则：

设 $B = \{+, -, *, /\}$ 是二元算术运算符的集合（也就是说，B 中的每个操作符 b 正好有两个操作数）。用 a 表示操作数。对于一个中缀表达式 I，用 I' 表示它的逆波兰式表达式的形式。然后：

（1）如果 I 是单个操作数 a，则逆波兰式表达式的形式为 a。

（2）如果 I1 b I2 是一个中缀表达式，其中 b 是 B 的一个元素，那么其对应的逆波兰式表达式是 I1' I2' b。

（3）如果（I）是一个中缀表达式，它的逆波兰式表达式的形式是 I'。

直接通过这些规则构造的递归算法，将在后面以函数（在数学意义上）的形式展示。在数学中，函数 F 应用于"参数"x，表示为 Fx 或 F(x)，返回 x 的函数值。例如，应用于参数 90（度）的三角函数 SIN，表示为 SIN90，返回值 1。将符号为 $\sqrt{}$ 的平方根函数应用于一个参数，比如 4（符号为 $\sqrt{4}$），返回值为 2。

相应的，这里命名为 RP，参数为中缀表达式 I 的算法如下：

RP (*I*)

Step 1: if *I* = *a* **then return** *a*

    else

Step 2:   **if** *I* = *I1 b I2*

      then return RP(*I1*)RP(*I2*)*b*

    else

Step 3:   **if** I = (*I1*) **then return** RP(*I1*)

Step 4: halt

步骤 3 中，一般形式的 **if** *C* **then** *S* 是 **if then else** 决策形式的一种特殊情况：只有当条件 *C* 为真时，控制流到 *S*，否则控制流到 **if then** 后面的语句。

因此，RP 函数可以使用"更小的"参数递归地激活自己。

很容易看出，RP是转换规则的直接实现，因此其在构造上是正确的。（并非所有算法都是如此不言而喻的正确；它们的理论基础可能要复杂得多，其正确性必须通过仔细的论证，甚至某种形式的数学证明；当然，也可能存在它们的理论基础薄弱，甚至根本不存在的情况。）

为了说明算法如何使用实际参数，请考虑以下示例：

(1) Suppose $I = a + b$. Then:

$\quad$ RP($a + b$) = RP(a) RP(b) +$\qquad$ (by step 2)

$\qquad$ = ab +$\qquad$ (by step 1 twice)

(2) Suppose $I = (a + b)*c$. Then:

$\quad$ RP($(a + b)*c$) = RP($a + b$) RP(c)*$\qquad$ (by step 2)

$\qquad$ = RP(a) RP(b) + RP(c)*$\qquad$ (by step 2)

$\qquad$ = $ab + c$*$\qquad$ (by step 1 thrice)

(3) Suppose $I = (a*b) + (c - 1/d)$. Then:

$\quad$ RP($(a*b) + (c - 1/d)$)

$\qquad$ = RP($a*b$) RP($c - 1/d$) +$\qquad$ (by step 2)

$\qquad$ = RP(a) RP(b)*RP(c) RP($1/d$) − +$\qquad$ (by step 2 thrice)

$\qquad$ = RP(a) RP(b)*RP(c) RP(1) RP(d)/ − +$\qquad$ (by step 2)

$\qquad$ = $ab*c1d/ − $ +$\qquad$ (by step 1 five times)

算法作为实用人工制品的"优度"

如前所述，仅仅设计一个正确的算法是不够的。就像任何实用人工制品的设计者一样，算法设计者必须关心算法有多好，它的工作有多高效。我们能从这个意义上衡量算法的好坏吗？我们可以用某种定量方式来比较完成相同任务的两种竞争的算法吗？

最明显的优点是算法执行所需的时间。但算法是一个抽象的人工制品，我们无法用物理时间来测量它；我们无法在真实的时钟上测量时间，因为算法本身并不涉及任何实质性的事物。假设笔者作为一个人执行一个算法，笔者认为可以计算在心里执行这个算法所花费的时间（也许可以借助铅笔和纸）。但这只是笔者对一组特定输入数据的算法性能的度量。我们关心的是在所有可能的输入中度量算法的性能，而不管谁在执行算法。

相反，算法设计者假设算法的每个基本步骤都需要相同的时间单位，将其视为"抽象时间"。他们根据问题所涉及数据项的数量来设计算法，从而构想出问题的规模。然后，他们采用两种算法优度的度量标准。一个与算法的最坏情况性能有关，它是问题规模 n 的函数；另一种方法处理它的平均性能，同样是问题规模 n 的函数。它们统称为时间复杂度。另一种度量方法是空间复杂度：执行算法所需的（抽象）存储空间量。

平均时间复杂度是更现实的优度度量，但它需要使用概率，因此更难分析。在本讨论中，我们将只处理最坏的情况。

考虑下面的问题。我有一个包含 n 个元素的列表。每一项包括一个学生的名字和他/她的电子邮件地址。列表按名字的字母顺序排列。我的问题是搜索列表并找到特定名字的电子邮件地址。

最简单的方法是从列表的开头开始，将每个条目的每个名称部分与给定的学生名字进行比较，沿着列表逐个进行，直到找到匹配，然后输出相应的电子邮件地址。（为了简单起见，我们假设学生的名字在列表的某个地方。）我们称之为"线性搜索算法"。

线性搜索

Input：student：n个条目的数组，每个条目由两个"字段"组成，分别表示名称（一个字符串）和电子邮件（一个字符串）。对于student中的第i个条目，用student[i].name和student[i].email表示各自的字段。

Input: given-name: 该名字正被"查找"

Temp variable i: 一个整数

Step 1: i←1;

Step 2: while given-name ≠ student[i].name

Step 3:　　　　do i←I + 1;

Step 4: output student[i].email

Step 5: halt

这里，**while** C **do** S的通用表示法指定了另一种形式的迭代：当条件C为真时，重复执行语句（"循环体"）S。与 **repeat** S **until** C相比，在每次迭代进入循环体之前，都要测试循环条件。

在最坏的情况下，期望的答案出现在最后一个（第n个）条目中。所以，在最坏的情况下，**while** 循环将被迭代n次。在这个问题中，列表中学生的数量n是关键因素，这是问题规模。

假设每一步所花费的时间大致相同。在最坏的情况下，该算法需要2n + 3个时间步才能找到匹配项。假设n非常大（比如20000）。在这种情况下，额外的因子"3"可以忽略不计。尽管将n加倍，但乘法因子"2"是一个常数因子。占主导地位的是n，即问题规模；这一点可能会因学生名单而异。我们感兴趣的是，用执行算法所需的（抽象的）时间作为这个n的函数，来描述这个算法的优度。

如果一个算法在时间 kn 内处理一个规模为 n 的问题，其中 k 是一个常数，我们说时间复杂度是 n 阶的，记为 O(n)。这称为大 O 表示法，由德国数学家保罗·巴赫曼在 1892 年引入。这种表示法为我们提供了一种将算法的效率（复杂度）指定为问题规模函数的方法。在线性搜索算法的情况下，其最坏情况的时间复杂度为 O(n)。如果一个算法在时间为 kn^2 的最坏情况下解决了一个问题，那么它的最坏情况时间复杂度为 O(n^2)。如果一个算法花费的时间是 k$n\log n$，那么它的时间复杂度是 O($n\log n$)，依此类推。

显然，对于规模为 n 的相同问题，O($\log n$) 算法需要的时间比 O(n) 算法少，O(n) 算法需要的时间比 O($n\log n$) 算法少，而 O($n\log n$) 算法需要的时间比 O(n^2) 算法少；后者将优于 O(n^3) 算法。最糟糕的算法是那些时间复杂度是 n 的指数函数的算法，比如 O(2^n) 算法。计算机科学家阿尔弗雷德·阿霍、约翰·霍普克罗夫特和杰弗里·乌尔曼在他们颇具影响力的著作《算法的设计与分析》(1974) 中，对具有这些时间复杂度的算法的优劣差异进行了鲜明的阐述。他们表明，假设有一定的物理时间来执行一个算法的步骤，O(n) 算法可以在一分钟内解决规模为 $n = 6*10^4$ 的问题；对于同样的问题，O($n\log n$) 算法可以解决规模为 $n = 4893$ 的问题；O(n^3) 算法解决了规模仅为 $n = 39$ 的同样问题；而 O(2^n) 的指数算法只能解决规模为 $n = 15$ 的问题。

因此，算法可以根据它们的大 O 时间复杂度进行分级，其中 O(k) 算法（k 是一个常数）在分级中最高，O(k^n) 指数算法最低。算法的优度会随着等级的下降而显著下降。

考虑学生列表搜索问题,但是这次要考虑到列表中的条目是按学生姓名的字母顺序排列的。在这种情况下,我们可以做我们在搜索电话簿或查阅字典时大致做的事情。当我们查找一个目录时,我们不会从第一页开始,一次一个地查找每个名称。相反,假设我们在字典中寻找其含义的单词以K开头,我们将字典的页面翻到K附近的页面。例如,如果我们打开字典到M附近,此时我们知道我们必须翻转回来;如果我们在H开头,我们就得继续向前翻。利用字母排序的优势,我们减少了搜索量。

这种方法可以更精确地遵循一种叫作二分搜索的算法。假设列表有 $k = 2^n - 1$ 个条目,在每一步中,中间的条目被标识。如果这样识别的学生姓名在字母顺序上比给定的名字"低",算法将忽略中间元素左侧的条目。然后,它将识别列表右半部分的中间条目,并再次进行比较。在每次查找中,如果没有找到名字,它将再次将列表减半,直到找到匹配的名字。

假设列表中有 $k = 15$(即 $2^4 - 1$)个条目。假设这些编号从1到15。那么很容易就可以确认算法经过的最大路径将是以下之一:

$8 \to 4 \to 2 \to 1$

$8 \to 4 \to 2 \to 3$

$8 \to 4 \to 6 \to 5$

$8 \to 4 \to 6 \to 7$

$8 \to 12 \to 10 \to 9$

$8 \to 12 \to 10 \to 11$

$8 \to 12 \to 14 \to 13$

$8 \to 12 \to 14 \to 15$

在这里，列表条目8是中间的项。因此，在找到匹配项之前，最多只搜索4 = $\log_2 16$项。对于规模为 n 的列表，二分搜索的最坏情况性能是 O（log n），这是对线性搜索算法的改进。

算法的美学

审美体验——对美的追求——不仅存在于艺术、音乐、电影和文学中，也存在于科学、数学甚至技术中。"美即是真理，真理之美"，这是约翰·济慈《希腊古瓮颂》（1820）最后几行诗的开头。英国数学家戈弗雷·哈罗德·哈代呼应济慈，彻底否定了"丑陋数学"的观点。

考虑一下为什么数学家要为某些特定的定理寻找不同的证明。一旦有人发现了一个定理的证明，为什么还要费劲去寻找另一个不同的证明呢？

答案是，当现有的定理在美学上没有吸引力时，数学家们会寻找新的定理证明，他们在数学中寻找美。

这同样适用于算法的设计。一个给定的问题可以通过某种算法来解决，该算法在某种程度上是丑陋的，即笨拙或缓慢的。有时这表现在算法的效率低下。因此，计算机科学家，尤其是那些受过数学训练的人，在算法中寻找美，就像数学家在证明中寻找美一样。也许最有说服力的算法美学发言人是来自荷兰的计算机科学家艾兹格·迪杰斯特拉、来自英国的C.A.R.霍尔和来自美国的高德纳。正如迪杰斯特拉曾经说过的，"美是我们的事业"。

这种审美欲望可以通过寻找更简单、结构更合理或使用"深度"概念的算法来满足。

例如，考虑本章前面描述的阶乘算法。该迭代算法基于阶乘函数的定义为：

$$\text{fact}(n) = 1 \text{ for } n = 0 \text{ or } n = 1$$
$$= n(n-1)(n-2)...3.2.1 \text{ for } n > 1$$

但是，阶乘函数有一个递归的定义：

$$\text{fact}(n) = 1 \text{ for } n = 0 \text{ or } n = 1$$
$$= n*\text{fact}(n-1) \text{ for } n > 1$$

对应的算法作为函数表示为：

rec-fact(n)
  if $n = 0$ **or** $n = 1$
    then return 1
    else return $n*\text{rec-fact}(n-1)$

许多计算机科学家会发现，这是一种更有美感的算法，因为它简洁、易于理解以及简朴的形式，而且它利用了阶乘函数更微妙的递归定义。请注意，递归和非递归（迭代）算法处于不同的抽象级别：递归版本可能由非递归版本的某些变体实现。

棘手（"非常困难"）的问题

在本章的最后，我将重点从解决计算问题的算法转移到计算问题本身。在前面的部分中,我们看到算法的性能可以根据它们的时间(或空间)复杂度来估计。例如,在学生列表搜索问题中,两种算法(线性搜索和二元搜索),尽管解决同一问题,但表现出两种不同的最坏情况时间复杂度。

但是,考虑所谓的"旅行推销员问题":给定一组城市和它们之间的道路距离,推销员是否可以从他的基础城市出发,访问所有的城市,然后返回到他的原点,这样旅行的总距离小于或等于某个特定的值?事实上,对于某些常数k和问题规模n(如城市数量),这个问题没有已知的算法小于指数时间复杂度$O(k^n)$。

如果解决问题的所有已知算法的时间复杂度至少是指数级的,那么计算问题就被称为棘手的——"非常困难"。存在多项式时间复杂度算法的问题——如$O(n^k)$——被认为是可处理的,即"实际可行的"。

处理计算问题的难易处理性的计算机科学分支是一个正式的数学领域,称为计算复杂性理论,在20世纪60年代和70年代初主要由以色列的迈克尔·拉宾、加拿大的斯蒂芬·库克和美国的尤里斯·哈特马尼斯、理查德·斯特恩斯和理查德·卡普创立。

复杂性理论家将两类问题分别称为**P**和**NP**。这些类的正式定义(即数学定义)是令人生畏的,与自动机理论有关,特别是与某些类型的图灵机(参见第二章),这里不占用篇幅展开介绍。非正式地来说,**P**类包含了所有可在多项式时间内解决的问题,因此这些问题是可处理的。**NP**类由一些问题组成,为这些问题提出的解决方案可能在多项式时间内得到,也可能不在多项式时间内得到,但可以在多项式时间内被检验其正确性。例如，旅行推销员问题没有一个（已知的）多项式时间算法解决方案，但给出了一个解决方案，它可以"很容易"地在多项式时间内检查解决方案是否正确。

但是，如前所述，旅行推销员的问题是棘手的。因此，NP类可能包含被认为是棘手的问题——尽管NP也包含P类易处理的问题。

这些想法的影响是相当大的。特别值得注意的是一个被称为NP完备的概念。如果π在NP中，并且NP中的所有问题都可以在多项式时间内转化或归约为π，则称问题π是NP完备的。这意味着，如果π是难以处理的，那么NP中的所有其他问题都是难以处理的。相反，如果π是可处理的，那么NP中的所有其他问题也都是可处理的。因此，在这个意义上，NP中的所有问题都是"等价的"。

1971年，斯蒂芬·库克提出了NP完备的概念，并证明了一个被称为"可满足性问题"的特殊问题是NP完备的。可满足性问题涉及布尔（或逻辑）表达式——例如表达式（a或b）和c，其中术语a、b和c是布尔（逻辑）变量，只有可能的（真）值TRUE和FALSE。问题是："是否有一组真值的条件，使得布尔表达式的值是TRUE？"库克证明了NP中的任何问题都可以简化为可满足性问题，该问题也存在于NP。因此，如果可满足性问题是易处理的，那么NP中的所有其他问题也是如此。

这就提出了以下问题：所有NP问题都有多项式时间算法吗？我们之前注意到NP包含P，但这个问题问的是：P是否与NP相同？这就是所谓的P = NP问题，可以说是理论计算机科学中最著名的开放问题。没有人证明P = NP，而且普遍认为事实并非如此；也就是说，人们普遍认为（但尚未证明）$P \neq NP$。这意味着 NP 中存在一些 P 中没有的问题（例如旅行推销员和可满足性问题），因此本质上是难以处理的；如果它们是 NP 完备的，那么所有其他可归约为它们的问题也是难以解决的。对于这类问题，没有实际可行的算法。

NP 完备理论告诉我们，许多看似不同的问题以一种奇怪的方式联系在一起。一个可以变成另一个，它们彼此等价。一旦我们意识到，在商业、管理、工业和技术中应用的大量计算问题（"现实世界"的问题）是 NP 完备的，我们就会理解这个想法的重要性：如果一个人能够为其中一个问题构建一个可行的（多项式时间）算法，那么他就可以为所有其他问题找到一个可行的算法。

那么，如何应对这些棘手的问题呢？一种常见的方法是第六章的主题。

第四章

编程的艺术、科学和工程

  重复一遍,算法思维是计算机科学的核心。然而,算法是抽象的人工制品。计算机科学家可以心满意足地(如果他们愿意的话)生活在算法的稀有世界中,而不会像"纯粹的"数学家那样冒险进入"现实世界"。但是,如果我们想要真正的物理计算机代表我们进行计算,如果我们想让物理计算机不仅完成那些我们认为过于烦琐(尽管必要)的计算,而且还要完成那些超出我们正常认知能力的计算,那么仅靠算法思维是不够的。它们必须以一种可以与物理计算机通信、解释和执行的形式来实现,而不是以人类的方式。

  这就是编程进入计算领域的地方。计算机程序是用一种可传递给物理计算机的语言对所需的计算进行说明的程序。构造这种计算的行为被称为程序设计,而用于指定程序的语言被称为程序设计语言。

程序是阈限人工制品

  程序的概念是难以捉摸、微妙且相当奇怪的。首先,正如我简要解释的那样，相同的计算可以在几个抽象级别上描述，具体取决于表示它的语言，从而允许多个等效的程序。其次，程序是双面的。一方面，程序是一段静态文本，即具有抽象人工制品所有特征的符号结构；另一方面，程序是一个动态的过程——也就是说，它在物理计算机中引起一些事情的发生，这些过程消耗物理时间和物理空间。因此，它有一个材料基础，并作用其上。此外，它还需要一种物质介质才能工作。

因此，程序有一个抽象的和物质的面貌，为此，我们可以称程序为阈限人工制品。这种阈限性的后果是巨大的，也是有争议的。

首先，在计算机科学界，一些人被程序的抽象性所吸引，他们认为程序是数学对象。对他们来说，编程是一种涉及公理、定义、定理和证明的数学活动。其他计算机科学家坚持其物质方面，并认为程序是经验对象。对他们来说，编程是一种经验的工程活动，涉及需求规范、设计、实现和进行测试并评估结果工件的实验等经典工程任务。

其次，程序和思维之间的类比也经常被描绘。如果程序是一个阈限的人造对象，那么思维就是一个阈限的自然对象。一方面，记忆、思考、感知、计划、语言理解和掌握等心理（或认知）过程可以（几个世纪以来一直如此）被研究，就好像思维是一个纯粹抽象的东西，与"真实"世界自主互动。然而，思维是有"座位"的。除非一个人是一个不悔改的二元论者，将思维与身体完全分离，否则你不会相信思维可以存在于大脑之外——一个物理对象。因此，当一些哲学家和认知科学家把思维当作一个抽象实体来研究时，神经科学家则从大脑过程的角度寻求对精神现象的严格物理解释。

事实上,认知的科学研究受到了思维与程序类比的深刻影响。其中一个结果就是计算机科学的一个分支——人工智能的发展,它试图创造出类似思维和大脑的计算人工制品。另一个是认知心理学转变为一门更广泛的学科,称为认知科学,其核心是假设心理过程可以建模为类似程序的计算。

因此,计算机科学的智力影响已经远远超出了学科本身。正如达尔文的进化论已经扩展到生物学以外的领域一样,由于思维与程序的类比,计算机科学的影响已经超出了计算本身。或者更确切地说(我们将在后面的章节中看到),计算的概念已经远远超出了物理计算机和自动计算的范围。我认为可以公平地说,很少有人工科学在它们自己的领域之外产生这样的智力影响。

然而,程序阈限性的另一个后果是,程序和程序设计与被称为编程语言的人工语言不可避免地交织在一起。人们可以使用自然语言来设计算法,也许还可以添加一些人工符号(就像在第三章中介绍的算法那样)。但是,没有掌握至少一种编程语言的人是不可能成为程序员的。一个简洁(如果只是粗略的话)的公式很可能是:

$$算法 + 编程语言 = 程序$$

这本身有几个含义。

一个是编程语言理论作为计算机科学的一个分支的发展。这不可避免地导致了这一理论与研究自然语言结构的语言学之间的关系。

第二个含义是对"梦想"编程语言的永无止境的追求，它可以比任何语言的前辈或当代竞争对手做得更好。这就是语言设计的活动。语言设计者面临的挑战是双重的：促进计算与物理计算机的通信，然后物理计算机可以在最少人工干预下执行这些计算；并促进与其他人的交流，以便他们能够理解计算、分析计算、批评计算并提出改进建议，就像人们处理任何文本一样。这种双重挑战一直是计算机科学家对编程和其他计算语言痴迷的根源。

与语言设计密切相关的是第三种成果：研究和开发被称为编译器的程序。它可以将用编程语言编写的程序翻译成特定物理计算机的机器代码。编译器设计和实现是计算机科学的另一个分支。

最后，人们还在努力设计物理计算机的特性，以促进编译器的任务。这种活动被称为"面向语言的计算机设计"，在计算机科学的分支——计算机体系结构中一直引起极大的兴趣（参见第五章）。

图3示意性地显示了程序和编程与这些其他实体和学科的许多关系。矩形的实体是计算机科学之外的学科；椭圆中的实体是计算机科学中的学科。

语言、思想、现实和编程

注意，我在前面说的是"语言"，而不是"符号"。符号指的是为书写某物而发明的符号，比如化学或数学符号。语言超越了符号，因为它为思考某事提供了一个符号系统，语言本身就夹杂着思想。

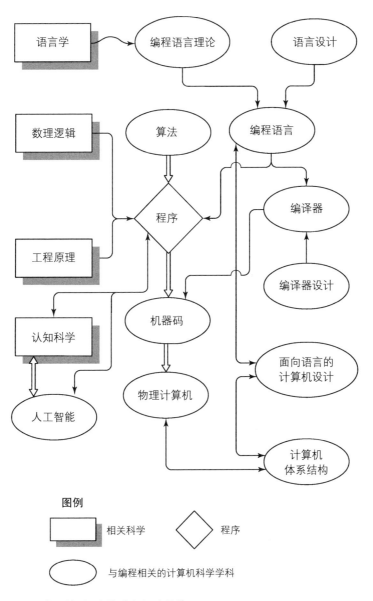

图3 程序设计、相关学科和相关科学

有一个著名的问题，语言学家和人类学家一直争论不休：思想是由语言定义的吗？有些人断言，事实就是如此，我们使用的语言决定了我们思考世界的方式和我们的想法，事实上，它定义了我们对现实本身的概念化。它最极端的含义是，由于语言是文化的决定性因素，思想、感知或概念不能从一种语言文化翻译到另一种语言文化。我们每个人都被困在自己语言定义的现实中。这是一个非常后现代的情况。其他人则持较为温和的观点，他们认为语言会影响但不会定义我们对世界的看法。语言定义或影响思想的命题被称为萨皮尔-沃尔夫假说，以人类语言学家爱德华·萨皮尔和本杰明·李·沃尔夫的名字命名。（这也被称为"语言相对性原则"。）

萨皮尔-沃尔夫假说研究的是自然语言。我们关注的是人工语言，特别是那些用来表达计算的语言。据我所知，没有人为计算领域制定过与萨皮尔-沃尔夫假说类似的框架，但是，计算机科学家从最早的电子计算机使用编程和其他计算语言起就表现出的痴迷则强烈表明，计算机科学界已经默认了这种假说的某种形式。更准确地讲，我们可以有信心地说，计算语言（特别是编程）与计算环境的性质紧密地交织在一起，并且编程语言会影响程序员的思维方式。

因此，让我们首先考虑编程语言。程序的本质将从这个讨论中自然地显现出来。

编程语言是抽象的人工制品

前面提到过，计算可以被指定为不同抽象级别的程序，这些程序与可以执行这些程序的物理计算机的"距离"不同。相应地，编程语言可以被设想在不同的抽象级别上。粗略的二分法将高级语言与低级语言区分开来。前者使程序可以独立于所有将要执行它们的物理计算机而编写，而后者指的是针对特定计算机系列，甚至更具体地针对特定计算机而设计的语言。

因此，高级语言是"与机器无关的"，而低级语言是"与机器相关的"，但需要注意的是，独立或依赖的程度可能会有所不同。最低级别的语言被称为汇编语言，它们特定于具体（家族或个人）的物理计算机，以致汇编语言程序员实际上操纵着计算机本身的特性。

在计算机历史上有一段时间，几乎所有的编程都是使用汇编语言完成的。这些程序仍然是符号结构（在相当程度上是抽象的），但被称为"汇编程序"的翻译器（它们本身的程序）会把它们转换成目标计算机的机器代码。然而，由于汇编语言编程的乏味、困难并且需要大量的人力时间和易出错性，人们的注意力转移到了越来越高级的、与机器无关的编程语言的发明和设计上，而将这些语言中的程序翻译成特定计算机的可执行机器代码的任务则委托给了编译器。

在本章以后以及本书的其余部分，除非明确说明，否则术语"编程语言"将始终指代高级语言。

与自然语言不同，编程语言是发明或设计出来的。因此，它们是人工制品。它们需要使用符号作为标志。正如我们将看到的，编程语言实际上是一组符号结构，并且独立于物理计算机，它们是抽象的，就像算法是抽象的一样。因此，我们遇到了这样一种奇怪的情况：虽然用这种语言编写的程序是阈限的，但编程语言本身是抽象的。

语言 = 符号 + 概念和类别

让我们重新审视符号和语言的区别。计算世界由数百种计算语言组成。这些语言大部分用于编程,但也有一些是为其他目的而创建的,特别是用于不同抽象级别的物理计算机的设计和描述(参见第五章)。后者通常被称为"计算机硬件描述语言"(CHDLs)或"计算机设计和描述语言"(CDDLs)。

这些计算语言使用不同的标记或符号,而学习一门新的计算语言的一部分脑力工作就是掌握符号,也就是符号所象征的东西。这需要将符号映射到基本的计算概念和语言类别上。因此,一种特定的语言由一系列概念和类别以及表示它们的符号组成。同样,用一个粗略的公式表示:

$$概念/类别 + 符号 = 语言$$

在计算中,不同的符号被用在不同的语言中来表示同一个概念。反而言之,同一个符号在不同的语言中可能象征着不同的概念。

例如,被称为赋值的基本编程概念(在第三章中已经遇到)在不同的语言中可以用" = >"、" = "、"←"、": = "等符号表示。因此,赋值语句可以表示为:

$$X + 1 = > X$$
$$X = X + 1$$
$$X: = X + 1$$
$$X \leftarrow X + 1$$
$$X + = 1$$

这些赋值语句表达的含义都是一样的：变量X的当前值加1,结果赋值给（或复制回）X。赋值是一个计算概念,赋值语句属于语言范畴,存在于大多数编程语言中。根据语言设计者的品味和偏好,不同的语言表达它的方式也不同。

编程语言中的概念和类别

那么这些概念和类别是什么呢？计算,召回,就是符号处理；用更通俗的说法,就是信息处理。"一开始是信息",或者像语言设计者喜欢说的那样,是要处理的数据。因此,嵌入在所有编程语言中的一个基本概念就是所谓的数据类型。如第一章所述,数据类型定义了数据对象（也称为"变量"）可以保存的值的性质,以及对这些值允许的（"合法"）操作。数据类型是"原始的"（或"原子的"）或"结构化的"（或"复合的"）,它们由更基本的数据类型组成。

在第三章的阶乘示例中,只有原始数据类型"非负整数",这意味着大于或等于0的整数可以作为该类型变量的值；对于这种类型的变量,只能执行整数算术运算。这还意味着只能将整数值赋给这种类型的变量。例如,这样的赋值：

$$x \leftarrow x + 1$$

如果x定义为整数类型,则为合法语句。如果x没有被这样声明,而是被声明为（比如）一个字符串（代表一个名字）,那么赋值将是非法的。

数字本身不一定是整数,除非它被定义为整数。因此,从计算的角度来看,电话号码不是整数,它是一个数字字符串。两个电话号码不能相加或相乘。因此，如果x被声明为一个数字字符串，先前的赋值将是无效的。

第三章的线性搜索算法包括基本数据类型和结构化数据类型。变量i的基本类型是整数；给定的变量为结构化类型"字符串"，其本身由原始类型"字符"组成。学生列表也是一种结构化数据类型，有时被称为"线性列表"，有时被称为"数组"。学生的第i个元素本身就是一种结构化数据类型，在不同的编程语言中有不同的名称，包括"记录"和"元组"。该元素由两种数据类型组成，其中，学生的名字（name）是字符串类型，另一个邮件信息（email）也是字符串类型。因此，像学生（student）这样的变量是一个分层组织的数据结构：字符组成字符串，字符串组成元组或记录；元组组成列表或数组。

但数据或信息只是计算的开始。此外，变量本身是被动的。计算包括操作以及将操作组合成过程。因此，编程语言不仅必须具有指定数据对象的功能，还必须包括指定操作和过程的语句。

事实上，在前面的章节中，我们已经多次遇到最基本的语句类型。一种是赋值语句。它的执行调用一个同时涉及方向和时间信息流的过程。例如，在执行赋值语句：

$$A \leftarrow B$$

当A和B都是变量，信息从B流到A。但它不像水从一个容器流到另一个容器。执行该语句后，B的值不会改变、减少或清空。相反，B的值被"读取"并"复制"到A中，因此在最后，A和B的值是相等的。但是，在执行语句

$$A \leftarrow A + B$$

A的值确实发生了变化：A的新值 = A的旧值 + B的值。其中B的值保持不变。

赋值语句的一般形式是

$$X \leftarrow E$$

其中，E是一个表达式，比如算术表达式 Y + 1，(X − Y) ∗ (Z/W)。赋值的执行通常是一个两步过程：首先计算E；然后这个值被赋给X。

然后，赋值语句指定计算中的操作单元。它是计算的原子过程。但就像在自然界中原子结合成分子，分子结合成更大的分子一样，在计算世界中也是如此。赋值语句合并形成更大的段，后者合并形成更大的段，直到获得完整的程序。计算世界就像自然界一样，存在着层次结构。

因此，计算机科学家的一项主要任务是发现组合规则，发明表示这些规则的语句类型，并为每种语句类型设计符号。虽然组合和语句类型的规则可能是非常通用的，但不同的编程语言可能使用不同的符号来表示它们。

其中一种语句类型是顺序语句：两个或多个（更简单的）语句按顺序组合，以便在执行时按照组件语句的顺序执行。在第三章中使用的表示序列的符号是";"。因此，在欧几里得算法中，我们得到顺序语句：

$m \leftarrow n$;
$n \leftarrow r$;
goto step 1

其中，控制流按照所示的顺序通过三个语句。

但计算过程也可能需要在多个备选中做出选择。在第三章的几个算法中使用的 **if ... then ... else** 语句是编程语言中条件语句类型的实例。在一般形式中，**if** *C* **then** *S1* **else** *S2*，条件 *C* 被评估，如果为真，则控制转到 *S1*，否则控制流到 *S2*。

有时，我们需要将控制流返回到较早的计算部分并重复它。在第三章的线性搜索和非递归阶乘算法中使用的 **while ... do** 和 **repeat ... until** 语句就是迭代语句类型的例子。

这三种语句类型——顺序语句、条件语句和迭代语句——是构建程序的基石。每种编程语言都提供了表示这些类别的符号。事实上，从原则上考虑，任何计算都可以由仅涉及这三种语句类型组合的程序指定。1966年，两位意大利计算机理论家科拉多·伯姆和朱塞佩·亚科皮尼证实了这一点。在实践中，已经提出了许多其他的组合规则和相应的语句类型来促进编程（例如在第三章欧几里得算法中使用的 **goto** 语句所代表的无条件分支）。

编程的艺术

编程是一种设计行为，与所有的设计活动一样，它需要判断、直觉、审美品味和经验。正是因为这一点，高德纳将他著名且颇具影响力的系列著作命名为《计算机程序设计艺术》（1968—1969）。大约十年后，高德纳在一次演讲中详细地阐述了这个主题。在谈到编程的艺术时，他暗指编程是一种艺术形式。程序应该在美学上令人满意，它们应该很漂亮。编写程序的体验应该类似于创作诗歌或音乐。因此，作为艺术、音乐和文学话语中如此密切的一部分,风格的概念必须成为编程美学的一个元素。

回想一下在第三章中提到的荷兰计算机科学家艾兹格·迪杰斯特拉的一句话:"在设计算法时,'美是我们的事业'。"俄罗斯计算机科学家安德烈·叶尔绍夫也认同这些观点。

围绕同样的主题,高德纳后来提出程序应该是文学作品,人们可以在编写程序时获得乐趣,这样程序就会给别人带来阅读的乐趣。(他称这种哲学为"文学编程",尽管我认为"文学编程"更适合表达他的情感。)

编程是一门数学科学

当然,计算机科学家(包括高德纳)寻求发现更客观和正式的编程基础,他们想要一门编程的科学。前面提到的伯姆-亚科皮尼结果,是计算机科学家渴望的那种形式的、数学的结果。事实上,许多计算机科学家所说的编程的"科学"指的是数学科学。

将程序设计视为一门数学科学的观点,在另外三个方面得到了最显著的体现——所有这些都与程序抽象的一面有关,或者,正如我在本章前面提到的,有一部分人认为程序是数学对象。

对编程科学的第一个贡献,是编程语言语法规则的发现。这些规则决定程序的语法正确性,并具有巨大的实际意义,因为编译器(将高级程序自动翻译成机器代码)的首要任务之一是确保其翻译的程序在语法或句法上是正确的。编程语言语法理论的起源,要归功于语言学家诺姆·乔姆斯基对自然语言语法理论的研究。

对编程科学的第二个贡献是语义规则的开发,即定义不同语句类型含义的原则。它的重要性应该是非常明显的:为了使用一种编程语言，程序员必须非常清楚其组件语句类型的含义。同样，编译器编写者必须清楚地理解每种语句类型的含义，才能将程序翻译成机器代码。但是语义，作为语言学中使用的术语，是一个棘手的问题，因为它涉及将语言类别与其在世界中所指代的事物联系起来，而编程语言语义学理论反映了这些相同的困难。可以公平地说，编程中的语义理论，尽管它的发展很成熟，但还没有被计算机科学界接受，也没有像语法理论那样有效地使用。

对编程科学的第三个贡献与语义问题密切相关。这一贡献是建立在诸如英国人C.A.R.霍尔和荷兰人艾兹格·迪杰斯特拉这样的计算机科学家的信念之上的，他们认为计算类似于数学，数学家使用的同样的原则，如公理、演绎逻辑规则、定理和证明，也适用于编程。1985年，霍尔明确而大胆地提出了这一理念，并发表了如下宣言：

（1）计算机是数学机器。也就是说，它们的行为可以从数学上定义，每个细节都可以从定义中逻辑地推导出来。

（2）程序是数学表达式。它们精确而详细地描述了执行它们的计算机的行为。

（3）编程语言是数学理论。它是一种形式化的系统，可以帮助程序员开发一个程序，并证明该程序满足其需求的规范。

（4）编程是一项数学活动。它的实践需要应用传统的数学理解和证明方法。

在古老的数学公理传统中，人们从公理（在相关领域被认为是"不言而喻"正确的命题，例如在第三章中提到的数学归纳法的原理）和基本概念的定义开始，利用演绎规则，从这些公理、定义和已经证明的定理中逐步证明新的见解和命题（统称为定理）。受这一传统的启发，对编程数学科学的第三个贡献，涉及基于公理、定义以及定义相关编程语言语义的演绎规则的程序正确性公理证明的构建。这种语义被称为公理语义，其应用被称为公理正确性证明。

就像数学中的公理化方法（以及它在数学、物理、经济学等学科中的应用）一样，在编程的数学科学中有很多形式上的优雅和美感。然而，公平地讲，虽然在这一领域已经发展出大量的知识体系，但许多学术计算机科学家和工业从业者仍然怀疑它们在喧嚣的计算"真实世界"中的实际适用性。

编程就是（软件）工程

这是因为许多人认为程序不仅仅是美的抽象人工制品。就连霍尔的宣言也承认，程序必须描述执行它们的计算机的行为。后者是程序的物质面，赋予它们阈限性。事实上，对许多人来说，程序是技术产品，因此编程是一项工程活动。

软件这个词似乎是在1960年就进入了计算机词汇表，但它的内涵仍然不确定。有些人使用"软件"和"程序"作为同义词；有些人认为软件是指一组特殊和必要的程序（如操作系统和其他工具以及统称为"系统程序"的"实用程序"），它们被构建为在物理计算机上执行以创建虚拟机（或计算机系统），其他人可以更有效地使用（参见第二章中的图1）。还有一些人认为软件不仅意味着程序，而且意味着对大型程序的开发、操作、维护和修改至关重要的相关文档。有些人会把人类的专业技术和知识包括在这个纲要里。

无论如何，"软件"有以下重要的含义：它是计算机系统的一部分，本身不是物理的；它需要物理计算机才能使其运行；从某种意义上说，软件在很大程度上是一种工业产品，就像这个形容词所暗示的那样。

因此，软件是一种计算人工制品，它促进了许多（可能是数百万，甚至数十亿）用户使用计算机系统。大多数时候（尽管并非总是如此），它是一种商业生产的人工制品，它体现了我们对工业系统所期望的一定程度的稳健性和可靠性。

也许与其他工业系统类似，软件开发项目与生命周期相关联。因此，像许多其他复杂的工程项目（例如新的太空卫星发射项目）一样，软件系统的开发被视为一个工程项目，在这种情况下，软件工程一词（在20世纪60年代中期首次出现）似乎特别合适。这自然就引出了"软件工程师"的概念。软件工程的很大一部分思想起源于工业部门，这并不是一个巧合。

在过去的五十年中，人们提出了各种软件生命周期模型。总的来说，他们都认识到软件系统的开发涉及若干阶段：

（1）软件用以服务的需求分析。

（2）从需求分析中识别的不同组成部分（"模块"）的精确功能、性能和成本规范的开发。

（3）（希望）满足规范的软件系统的设计。这项活动本身可能包括概念和详细的设计阶段。

（4）将设计实现为以编程语言指定并编译为在目标计算机系统上执行的操作软件系统。

（5）验证和确认已实施的一套方案，以确保它们符合规范。

（6）一经验证和确认，系统的维护以及在必要时的修改。

当然，这些阶段不会以严格的线性方式进行。如果发现了缺陷和故障，总是有可能从较后的阶段返回到较早的阶段。此外，这个软件生命周期还需要一个基础架构——工具、方法、文档标准和人类专业知识，这些共同构成了软件工程环境。

　　人们还必须注意到，这些阶段中的一个或多个将需要坚实的科学理论体系作为其应用的一部分。规范和设计可能涉及使用具有自己语法和语义的语言；详细的设计和实现将涉及编程语言，可能还涉及公理证明技术。验证和确认总是需要复杂的软件验证和实验测试模式。正如经典的工程领域（如结构或机械工程）需要工程科学作为组件一样，软件工程也是如此。

第五章

计算机体系结构的学科

物理计算机位于图1所示层级结构"我的电脑"的底部。在日常用语中,物理计算机被称为硬件。它之所以被称为"硬"的,是因为它是一个最终遵守自然法则的物理人工制品。物理计算机是计算机科学家感兴趣的基本材料计算人工制品。

但如果有人问:"物理计算机的本质是什么?"我的回答可能会含糊其辞。这是因为物理计算机虽然是更大层级结构的一部分,但它本身足够复杂,可以显示自己的内部层级结构。因此,可以在多个抽象级别上设计和描述它。这些层级之间的关系结合了第一章中讨论的组合、抽象/细化和构造层级的原则。

也许从计算机科学家的角度来看,这种层级结构最重要的方面是,物理计算机是符号处理的计算人工制品,而遵守物理定律的物理组件是实现这种人工制品。这个发现要归功于匈牙利裔美国数学家和科学界的"搅局者"约翰·冯·诺伊曼的天才,是他首先认识到这一点,这种分离是非常重要的。作为一个符号处理器,计算机是抽象的,就像软件是抽象的一样。然而,就像软件一样，这个抽象的人工制品如果没有其物理实现就不存在。

把物理计算机看作一个抽象的、符号处理的计算人工制品，这构成了计算机的体系结构。（我之前使用术语"体系结构"来表示图1中所示的虚拟机的功能结构。但现在，"计算机体系结构"有了更具体的技术内涵。）实现架构的物理（数字）组件——实际的硬件——构成了它的技术。因此，我们有了这种"计算机体系结构"和（数字）技术之间的区别。

这种区别还有另一个重要的方面，即可以使用不同的技术实现给定的体系结构。体系结构不是独立于技术的，因为后者的发展影响着结构设计，但计算机体系结构的设计者享有一定程度的自主权或"自由度"。反之，体系结构的设计可能会影响应用的技术类型。

打个比方，考虑一个机构，比如大学。这既有抽象的特征，也有物质的特征。大学的组织，它的各种行政和学术单位，它们的内部结构和功能等等，都类似于计算机的结构。人们可以设计一所大学（它毕竟是一件人工制品），描述它，讨论和分析它，批评它，改变它的结构，就像人们可以设计任何其他抽象实体一样。但是，大学是通过人力资源和物质资源来实现的。它们类似于计算机（硬件）技术。因此，虽然大学的设计或发展需要相当程度的自主权，但其实现只能取决于其资源的性质、可用性和有效性（雇员、建筑、设备、物理空间、校园整体结构等）。反过来说，大学组织的设计将影响必须到位的各种资源。

以下是本次讨论的关键术语：计算机体系结构是计算机科学中的一门学科，涉及对物理计算机的逻辑组织、行为和功能元素的设计、描述、分析和研究；所有这些构成了（物理）计算机的架构。计算机架构师的任务是设计架构,一方面满足物理计算机用户(软件工程师、程序员和算法设计人员、非技术用户)的需求,同时又在经济和技术上可行。

因此,计算机架构是阈限的人工制品。计算机架构师必须在计算机功能和性能需求以及技术可行性之间巧妙地穿梭。

唯我论和社交型计算机

一切事物的全球化在很大程度上都要归功于计算机。"没有人是一个自成一体的孤岛",如果这句话成立,那么在21世纪计算机也不是。但是在很久很久以前,计算机确实作为它们自己的孤岛而存在。图1中描述的文本类型的计算人工制品会执行它的任务,就好像外面的世界不存在一样。它与环境的唯一交互方式是输入数据和命令并输出结果。除此之外,出于所有实际目的,物理计算机,连同它的专用系统、应用程序以及其他工具(如编程语言)都活在辉煌的、唯我论的孤立之中。

但是,正如刚才提到的,现在很少有计算机是唯我论的。互联网的出现,电子邮件的建立,万维网和各种形式的社交媒体已经终结了计算唯我论。即使是最孤僻用户的笔记本电脑或智能手机,一旦用户上网购买一本书或查看天气情况,或寻找去某个地方的路线时,它就会成为一台社交电脑。他的计算机具有社交性,因为它通过互联网与遍布全球的无数其他计算机(尽管对他来说是幸福的未知)进行交互和通信。事实上,每一个电子邮件发送者、每一个信息搜寻者、每一个在线视频观看者,都不仅仅是在使用互联网,她的电脑也是互联网的一部分。或者更确切地说,互联网是一个由社交计算人工制品和人类代理组成的全球互动社群。

但还有一些更适度的网络,计算机可以成为其中的一部分。一个组织(如大学或公司)内的机器通过所谓的"局域网"相互连接。分布在一个区域上的计算机群可以协作,每个计算机执行自己的计算任务,但在需要的时候交换信息。这种系统传统上被称为多计算或分布式计算系统。

多计算或分布式计算或互联网计算的管理,受到被称为"协议"的网络原理和极其复杂的软件系统的组合影响。然而,无论计算机是唯我论的还是社交的,当我们考虑计算机体系结构学科时,引起我们注意的是个人计算机。这就是我在本章其余部分讨论的内容。

外部和内部系统结构

20世纪60年代初,国际商用机器公司(IBM)的三位工程师吉恩·阿姆达尔、弗雷德里克·布鲁克斯和格里特·布洛乌首次在计算机环境中使用了"体系结构"这个词。他们用这个术语来表示物理计算机的功能属性集合,这些功能属性可供计算机底层架构程序员(使用汇编语言构建操作系统、编译器和其他基本实用程序的系统程序员)使用。可以说,这是它的"外立面"。然而,从那时起,计算机体系结构的实践已经扩展到包括机器物理(硬件)组件的内部逻辑、结构、功能组织和行为。因此,在实践中,"计算机体系结构"指的是物理计算机外部和内部的功能和逻辑方面。然而,这两个方面并没有一致的条件。在这里,为了简单起见,我将它们分别称为"外部"和"内部"系统结构。

这两者是层级相关的。它们是物理计算机的两种不同的抽象，外部系统结构是内部系统结构的抽象，或者反过来说，内部系统结构是外部系统结构的细化。或者，根据所使用的设计策略，可以将计算机的内部系统结构视为外部系统结构的实现。

计算机外部系统结构的设计是由计算机的计算环境所施加的力形成的。由于外部系统结构是物理计算机和系统程序员之间的接口，而系统程序员创建了计算机的"普通"用户可以"看到"的虚拟机，因此这种环境所要求的功能需求自然会对外部架构设计产生影响。例如，如果计算机C旨在支持以特定类型语言（例如L）编写的程序的有效执行，那么C的外部系统结构可能面向L的特征，从而减轻语言编译器将以L编写的程序翻译成C的机器代码的任务。或者，如果位于C之上的操作系统OS具有某些功能，将适当的特性纳入C的外部架构中可能会促进操作系统的实现。

另一方面，由于计算机的内部系统结构将由物理（硬件）组件实现，而这些组件是使用特定类型技术T构建的，内部系统结构的设计将受到T特性的限制。

与此同时，外部系统结构的设计可能会受到内部系统结构性质的影响和约束，反之亦然。因此，计算环境、计算机的外部系统结构、内部系统结构和物理技术之间存在着密切的关系（图4）。

外部系统结构

计算机外部系统结构的密室是它的指令集，它规定了程序员可以直接命令计算机执行的所有操作。究竟可以执行何种类

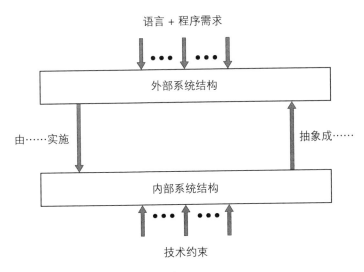

图4　计算机体系结构及其外部约束

型的操作,将由计算机直接支持或"识别"的数据类型决定。例如,如果计算机旨在有效地支持科学和工程计算,则其重要的数据类型将是实数(例如$6.483,4*10^8,-0.000021$等)和整数。因此,指令集应该包括一系列算术指令。

除了这些特定领域的指令之外,总会有一系列通用指令,比如用于实现条件(例如,**if then else**)、迭代(例如,**while do**)和无条件分支(例如,**goto**)等编程语言结构体。其他指令可以使程序被组织成负责不同类型计算的段或模块,并具有将控制从一个模块转移到另一个模块的能力。

指令实际上是一个"包",它描述了要执行的操作,以及对其输入数据对象(体系结构词汇表中的"操作数")的位置("地址")和输出数据放置位置的引用。这个地址的概念意味着一个存储空间。因此,存储器组件是外部系统结构的一部分。此

外,这些组件通常形成一个层级结构:

长期存储器,被称为"后备存储器"、"辅助存储器"或"硬盘驱动器"。

中期存储器,也被称为"主存储器"。

非常短期(工作)存储器,被称为寄存器。

这是一个在信息的保留性、大小、容量和访问速度方面的层级结构。因此,尽管外部架构是抽象的,但物理计算机的物质方面是可见的。空间(大小容量)和时间是物理的,用物理单位(信息的位或字节,时间的纳秒或皮秒等)测量,而不是抽象的表示。正是这种组合使计算机体系结构(外部或内部)成为阈限人工制品。

长期存储器是最持久的记忆("记住"的能力),从实际意义上说,它是永久的。它的容量是最大的,但访问速度也是最慢的。中期存储器只在计算机运行时保留信息。计算机断电后,这些信息就丢失了。它的大小容量远小于长期存储器,但访问时间远短于长期存储器。短期或工作存储器可能在一次计算的过程中多次改变其内容;它的容量比中期存储器低几个数量级,但访问时间比长期存储器短得多。

使用存储器层级结构的原因,是为了在存储保持与计算所需空间和时间之间保持明智的平衡。在指令集中,也会有指令来影响程序和数据在这些存储器组件之间的传输。

外部系统结构的其他特性,是围绕指令集及其数据类型集构建的。例如,指令必须有识别操作数和指令在存储器中位置("地址")的方式。识别存储器地址的不同方式被称为"寻址模式"。也会有规则或惯例来组织和编码各种类型的指令,以便它们可以有效地保存在存储器中。这样的约定被称为"指令格式"。同样,"数据格式"是组织各种数据类型的约定;特定数据类型的数据对象根据相关的数据格式保存在存储器中。

最后,一个重要的结构参数是字长。这决定了可以同时读取或写入中期(主)存储器的信息量(以位数衡量)。执行指令的速度在很大程度上取决于字长,以及单位时间内可访问的数据范围。

下面是一些用符号(汇编语言)表示法编写的计算机指令(或机器指令,之前已经使用过的术语)的一些典型示例,以及它们的语义(即这些指令导致发生的操作)。

指令	意义(作用)
1. LOAD R2,(R1,D)	R2←main-memory[R1+D]
2. ADD R2,1	R2←R2+1
3. JUMP R1,D	**goto** main-memory[R1+D]

说明:

R1,R2:寄存器

main-memory:中期(主)存储器

1:整数常量"1"

D:一个整数

在(1)中的"R1+D"将数字"D"添加到寄存器R1的内容中,从而确定了操作数的主存储器地址。(3)中的"R1+D"同样计算一个地址,但这个地址被解释为主存储器中一条指令的地址,并将控制权转移给该指令。

内部系统结构

一台物理计算机最终是由电路、电线和其他物理部件组成的综合体。原则上,外部系统结构可以解释为这些物理组件的结构和行为的结果。然而,像外部系统结构这样的抽象人工制品与物理电路之间的概念距离是如此之大,以至尝试这样的解释并不比从细胞生物学角度解释或描述一个完整的生物体(可能除了细菌和病毒)更有意义。比方说,细胞生物学不足以解释心血管系统的结构和功能。在了解整个系统之前,需要了解高于细胞水平的实体(如组织和器官)。同样地,无论是一台笔记本电脑还是世界上最强大的超级计算机,数字电路理论也不足以解释该计算机的外部系统结构。

在计算机中,这种概念上的距离,有时被称为"语义鸿沟",是以一种分层的方式弥合的。外部系统结构的实现,是根据内部系统结构及其组件来解释的。如果内部系统结构本身很复杂,而且与电路级别仍有概念上的距离,那么内部系统结构就可以用更低的抽象层次——我们称之为微架构——来描述和解释。微架构又可以被细化到所谓的"逻辑层",这可能足够接近电路层,从而可以用后者的组件来实现。一般来说,一台物理计算机可具有以下描述/抽象层次:

第4层:外部系统结构

第3层:内部系统结构

第2层:微架构

第1层:逻辑层

第0层:电路层

计算机架构师通常关心外部和内部系统结构,以及内部系统结构的细化,如前面所示的微架构(这种细化将在后面解释)。他们不仅对构成这些架构层级的特征感兴趣,而且对它们之间的关系感兴趣。

计算机内部系统结构的主要组件如图5所示,它包括以下内容。第一,存储器系统包括外部系统结构中可见的存储器层级结构,但包括只有内部系统结构级别可见的其他组件。这个系统包括负责管理信息(符号结构)的控制器,这些信息在层级结构的存储器之间,以及系统和计算机的其他部分之间传递。第二,一个或多个指令解释单元,它为执行准备指令并控制指令的执行。第三,一个或多个执行单元,负责实际执行计算中所需的各种指令。(指令解释系统和执行单元统称为计算机的处理器。)第四,用于在其他功能组件之间传递符号结构的通信网络。第五,输入/输出系统,负责从物理计算机环境接收符号结构,并

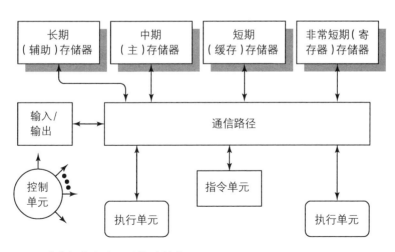

图5　计算机内部系统结构的描述

向物理计算机环境发送符号结构。最后,控制单元负责发出信号来控制其他组件的活动。

执行单元很像活体的器官。它们可以高度专用于对特定数据类型进行特定类型的操作,或者能够执行综合操作集的更通用目的单元。例如,一个执行单元可能只用于执行整数算术运算,而另一个则专用于实数的算术运算;第三个仅以各种方式操作位串;另一个对字符串进行操作,依此类推。

在处理器内部,它有自己专用的、极短期或瞬态的存储元件(比外部系统结构中可见的寄存器具有更短的保持性,有时称为"缓冲区寄存器"),在指令被指令解释或处理单元实际处理之前,必须从其他存储器中把信息带入其中。这样的缓冲区寄存器形成了内部系统结构中可见的存储层级中的"最低"层。

存储器层级结构的另一个组件在内部系统结构中可见,但(通常)在外部系统结构中被抽象出来。这是一个被称为高速缓存的存储器元素,位于中期(主)存储器和非常短期(寄存器)存储器之间。在图5中,这显示为"短期存储器"。它的容量和访问速度介于两者之间。缓存的基本思想是,由于程序模块中的指令(通常)按顺序执行,因此可以将一大块指令放在缓存中,这样可以比访问主存储器更快地访问指令。同样,程序行为的本质是,数据也经常从主存储器中的顺序地址访问,因此数据块也可以放置在高速缓存中。只有在缓存中没有找到相关指令或数据对象时,才会访问主存储器,这会导致缓存中的一块信息被相关信息所在的新块替换,以便以后在缓存中可以引用指令和数据。

"计算机中的计算机"

那么我们如何将外部系统结构连接到内部系统结构呢？它们实际上是如何关联的？要理解这一点，我们需要了解控制单元的功能（在图5中，控制单元位于一个漂亮的黑盒子般的隔离环境中）。

控制单元好比是计算机的大脑，是一种人造人，有时被描述为"计算机中的计算机"。它是管理、控制和排序其他系统所有活动以及它们之间符号结构运动的器官。在需要时，它通过向机器的其他部分发出控制信号（与指令和数据完全不同的符号结构）来实现这一点。就好比是木偶师拉动绳子来激活其他类似木偶的系统。

具体而言，控制单元向处理器（指令解释和执行单元的组合）发出信号，使处理器执行一种重复算法，通常称为指令周期（ICycle）。正是指令周期将外部系统结构与内部系统结构联系在一起。它的一般形式如下：

指令周期：

输入：主存储器,中期存储器；寄存器,短期存储器

内部：pc,瞬态缓冲区；ir,瞬态缓冲区；or,瞬态缓冲区

{pc,"程序计数器"的缩写,保存下一条要执行的指令的地址；ir,"指令寄存器",保存当前要执行的指令；or,保存指令操作数的值}。

取指令：使用pc将指令的值从主存储器转移到ir(ir←主存储器[pc])。

解码：对 ir 中指令的操作部分进行解码。

计算操作数地址：解码 ir 指令中操作数的地址模式，并确定操作数的有效地址和结果在主存储器或寄存器中的位置。

取操作数：将内存系统中的操作数提取到 or 中。

执行：使用 or 的操作数作为输入执行 ir 指令中指定的操作。

存储：在 ir 中指定的结果的目标位置存储操作的结果。

更新 PC：如果在"执行"步骤中执行的操作不是 **goto** 类型的操作，则 pc←pc + 1。否则什么都不做；"执行步骤"会将 ir 中的目标 **goto** 指令的地址放入 pc。

指令周期由控制单元控制，但它是指令解释单元，通过指令周期的取操作数步骤执行取指令，然后是存储和更新步骤，执行单元将实施执行步骤。作为一个具体的例子，考虑前面描述的加载（LOAD）指令：

LOAD R2,（R1,D）

请注意，这条指令的语义在外部系统结构层级只是

R2←主存储器[R1 + D]

在内部系统结构层级，它的执行需要指令周期的性能。指令被提取到 ir 中，它被解码，操作数地址被计算，操作数被提取，指令被执行，并且结果存储到寄存器 R2 中。对于外部系统结构的用户（系统程序员）而言，指令周期的所有这些步骤都在外部系统结构中被抽象为不必要的细节。

微程序

重复一遍，指令周期是一种算法，其步骤在控制单元的控制下。事实上，人们可能很容易理解，该算法可以实现为由控制单元执行的程序，而计算机的其余部分（存储器系统、指令解释单元、执行单元、通信路径、输入/输出系统）则作为"程序"环境的一部分。这一见解，以及基于这一见解设计的控制单元架构，被其发明者、英国计算机先驱莫里斯·威尔克斯命名为微程序设计。从某种意义上说，微程序控制单元执行一个微程序，该微程序为每种不同类型的指令实现指令周期，这导致一些人将微程序控制单元称为计算机中的计算机。事实上，微程序员所看到的计算机体系结构必然比图5中所示的内部系统结构更精细。这个微程序员（或控制单元实现者）对计算机的看法就是前面提到的"微架构"。

并行计算

制造人工制品的人（"工匠"）——工程师、艺术家、手工艺者、作家等——的天性，就是永远不会满足于他们所做的；他们渴望不断制造更好的人工制品（无论"更好"的标准是什么）。在物理计算机领域，两个主要的需求是空间和时间：制造更小更快的机器。

实现这些目标的一种策略是改进物理技术。这涉及固态物理学、电子学、制造技术和电路设计。自集成电路诞生以来，六十多年中取得了非凡的进步，生产出越来越密集、越来越小的组件，其中日益增长的计算能力集中在这些组件上，这对所有使用笔记本电脑、平板电脑和智能手机的人都是显而易见的。有一个著名的猜想被称为摩尔定律,以其发明者、美国工程师戈登·摩尔的名字命名,即单个芯片上的基本电路元件的密度大约每两年翻一番,这已在多年的经验中得到证实。

但是,考虑到物理技术的特定水平,计算机架构师已经发展出了提高计算吞吐量或速度的技术,例如,通过诸如单位时间处理的指令数量或单位时间内某些关键操作的数量等指标(例如在专用于科学或工程计算的计算机的实数算术运算)来衡量。这些体系结构策略属于并行处理的范畴。

基本的想法很简单。两个进程或任务T1、T2,如果发生在顺序任务流中(例如顺序程序中的指令)并且相互独立,则称它们是可并行执行的。如果它们满足某些特定条件,则可以实现这种相互独立性。这种情况的确切性质和复杂性将取决于以下几个因素:

(1) 任务的性质。

(2) 任务流的结构,例如它是否包含迭代(**while do**类型的任务)、条件(**if then elses**)或 **goto** 语句。

(3) 执行任务的单元的性质。

例如,考虑两个相同的处理器共享一个内存系统的情况。我们想知道在什么条件下,顺序任务流中出现的两个任务T1、T2可以并行启动。

假设T1的输入数据对象集和输出数据对象集分别指定为INPUT1和OUTPUT1。同样,对于T2,我们分别有INPUT2和OUTPUT2。假设这些输入和输出位于主存储器和/或寄存器中。如果满足以下所有条件,则T1和T2可以并行执行(记为

T1 ‖ T2）：

（1）INPUT1 和 OUTPUT2 是独立的（也就是说，它们没有任何共同点）。

（2）INPUT2 和 OUTPUT1 是独立的。

（3）OUTPUT1 和 OUTPUT2 是独立的。

这些被称为伯恩斯坦条件，以计算机科学家A.J.伯恩斯坦命名，他首次将其形式化。如果三个条件中的任何一个不满足，则它们之间就存在数据依赖关系，任务无法并行执行。作为一个简单的例子，考虑一个由以下赋值语句组成的程序流段：

[1] $A \leftarrow B + C$;
[2] $D \leftarrow B*F/E$;
[3] $X \leftarrow A - D$;
[4] $W \leftarrow B - F$.

由此：

INPUT1 = {B,C}, OUTPUT1 = {A}

INPUT2 = {B,E,F}, OUTPUT2 = {D}

INPUT3 = {A,D}, OUTPUT3 = {X}

INPUT4 = {B,F}, OUTPUT4 = {W}

应用伯恩斯坦条件，可以看出：（1）语句[1]和[2]可以并行执行；（2）语句[3]和[4]可以并行执行；但是，（3）语句[1]和[3]具有数据依赖性（变量A）；（4）语句[2]和[3]具有数据依赖性约束（变量D）；因此，这些语句对不能并行执行。

因此，实际上，考虑到这些并行和非并行条件，并假设有

97 足够多的处理器可以同时执行并行语句,语句的实际执行顺序将是:

语句[1]‖语句[2];

语句[3]‖语句[4]。

这种顺序/并行排序说明了并行程序的结构,其中任务是使用编程语言描述的单独语句。但是,考虑到物理计算机本身,并行处理研究的目标大致有两个:(1)发明能够检测任务之间并行性的算法或策略,并将并行任务调度或分配给计算机系统中不同的任务执行单元;(2)设计支持并行处理的计算机。

从计算机架构师的角度来看,并行性的潜力存在于多个抽象级别。其中一些并行级别是:

(1)任务(或指令)流在独立数据流、不同的多个处理器上同时执行,但任务流相互通信(例如,通过相互传递消息或相互传输数据)。

(2)任务(或指令)流在单个共享数据流、单个计算机内的多个处理器上并发执行。

(3)多个数据流占用多个存储单元,由在单个处理器上执行的单个任务(指令)流并发访问。

(4)单个任务流的段(称为"线程"),在单个处理器或多个处理器上同时执行。

98 (5)在指令周期内同时执行的单个指令的阶段或步骤。

(6)在计算机控制单元内同时执行微程序的部分。

所有并行处理架构都利用了刚才讨论的各种可能性,通常是组合使用。例如,考虑前面给出的抽象级别(5)。这里的想法是,由于指令周期由几个阶段组成(从取指令步骤到更新pc),因此执行指令周期的处理器本身可以由这些阶段组成的管道形式组织起来。一条指令将按顺序通过流水线的所有阶段。这个抽象层次的"任务"是指令在指令周期中移动的步骤。但是,当一条指令占据其中一个阶段时,其他阶段是空闲的,它们可以处理指令流中其他指令的相关阶段。理想情况下,一个七步指令周期可以由一个七阶段指令处理器流水线执行,并且流水线的所有阶段都很繁忙,以流水线的方式并行处理七条不同的指令。这当然是理想的条件。在实践中,指令流中的指令对可能违反伯恩斯坦条件,因此由于指令对各阶段之间的数据依赖性约束,流水线可能会有"空的"阶段。

支持这种并行处理的体系结构被称为流水线体系结构(图6)。

图6 一条指令流水线

作为另一个例子,考虑前面介绍的抽象级别(1)上的任务。在这里,同一台计算机内的多个处理器(也称为"内核")并行执行指令流(属于不同的程序模块)。这些处理器可能正在访问单个共享存储系统,或者存储系统本身可能被分解为不同的存储模块。无论如何,一个复杂的"处理器-存储器互连网络"(或"交换机")将充当存储系统和处理器之间的接口(图7)。这种方案被称为多处理器架构。

如前所述,并行处理体系架构的目标是,通过纯粹的体系架

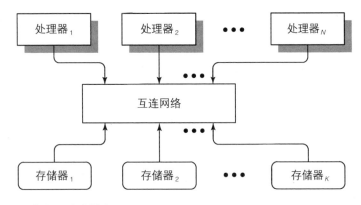

图7 多处理器的描述

构手段来增加计算机系统的吞吐量或速度。然而,正如四个赋值语句的小例子所示,由于数据依赖性约束,任务流的"并行化"存在限制。因此,在并行处理环境中可以实现的加速是有限度的。这个极限是由计算机设计师吉恩·阿姆达尔在20世纪60年代定量提出的,他指出并行处理计算机的潜在加速受到无法并行化的计算部分的限制。因此,增加并行执行单元数量的加速效果在某个点之后趋于平稳。这个原理被称为阿姆达尔定律。

计算机体系结构科学

读到本章这一部分的读者可能会问:既然计算机体系结构是阈限人工制品,那么从什么意义上说,这门学科是一门人工科学?

要回答这个问题,我们必须认识到该学科最引人注目的方面是它的知识空间(主要)由一系列启发式原则组成,而设计计算机体系结构中使用的推理就是启发式推理。

启发式——来自希腊词 hurisko,意为"寻找"——是为某些类型的问题(在下一章中讨论)提供解决方案的希望或承诺的规则或命题，但是没有成功的保证。套用匈牙利裔美国数学家乔治·波利亚的话说，他认识到启发式在数学发现中的作用，启发式思维从来都不是最终的、确定的、肯定的，相反，它是临时的、合理的、试探性的。

我们经常不得不使用启发式算法，因为我们可能没有任何其他选择。启发式是在缺乏更正式、更确定、基于理论的原则的情况下被调用的。第三章中讨论的分而治之原则是在问题解决和决策制定中普遍使用的启发式方法的一个例子。这是一个看似合理的原则，可能有助于解决复杂的问题，但不能保证在特定情况下成功。经验知识是许多启发式方法的来源。规则"如果天多云，带伞"就是一个例子。这把防护伞可能是合理的，但并非总是如此。

启发式的使用带来了实验的必要性。由于启发式不能保证成功，唯一的办法是将其应用于特定问题并凭经验查看它是否有效，也就是进行实验。反过来说，启发式原则本身可以基于先前的实验得出。启发式和实验齐头并进，启发式思维的先驱艾伦·纽厄尔和赫伯特·西蒙，以及计算机设计的先驱莫里斯·威尔克斯充分掌握了这一洞察力。

所有这一切都是以下内容的前奏：计算机体系结构学科是一门实验性的、启发式的人工科学。

自电子数字计算机问世以来的几十年里，关于计算机体系结构设计的规则、原则、规范、命题和模式已经形成，几乎所有这些在本质上都是启发式的。将存储层级结构作为设计原则的想法就是一个例子，流水线的原理是另一个例子。它们来源于经验知识、类比和常识观察。

例如，先前架构设计的经验和使用编译器生成机器代码所面临的困难，已经产生了消除这些困难的启发式原则。在20世纪80年代，计算机科学家威廉·伍尔夫根据为某些计算机设计编译器的经验提出了几种这样的启发式方法。这里是其中的一些：

（1）规律性。如果一个特定的（架构的）特性在体系结构的一个部分以某种方式实现，那么它应该在所有部分以相同的方式实现。

（2）关注点分离（分而治之）。整体架构应该可以划分为许多独立的功能，每个功能都可以单独设计。

（3）可组合性。凭借上述两个原则，应该可以任意方式组合单独的独立特征。

但实验必须遵循将启发式原则纳入设计。此类实验可能需要实施"原型"或实验机器并对其进行测试。或者，它可能涉及构建体系结构的（软件）仿真模型，并在仿真体系结构上进行实验。

在任何一种情况下，实验都可能揭示设计中的缺陷，在这种情况下，结果将是通过拒绝某些原则并插入其他原则来修改设计；然后重复实验、评估和修改的循环。

当然，这个模式几乎与科学哲学家卡尔·波普尔提出的科学问题解决模型相同：

$$P1 \rightarrow TT \rightarrow EE \rightarrow P2$$

这里，P1是"当前"的问题情况；TT是为了解释或解决问题情境而提出的一种尝试性理论；EE是应用于TT的错误消除过程（通过实验和/或批判性推理）；P2是消除错误后产生的新问题情况。在计算机体系结构的背景中，P1是设计问题，规定了最终计算机必须满足的目标和要求；TT是基于启发式的设计本身（这是一种计算机理论）；EE是对设计进行实验和评估，并消除其缺陷和局限性的过程；结果P2是一组可能修改的目标和要求，构成一个新的设计问题。

103

第六章

启发式计算

首先,许多问题并不利于用算法解决。教孩子骑自行车的父母不能给孩子展示一个他可以学习的算法,并像他学习两个数字相乘那样应用它。创意写作或绘画老师不能为学生提供写魔幻现实主义小说或绘制抽象表现主义画布的算法。

这种无能,部分是由于一个人(甚至是一个创造性写作的教员)的无知,或是对此类任务的确切性质缺乏了解。画家想要捕捉天鹅绒长袍的质感、苹果的坚固、微笑的神秘。但是,从绘画的角度来看,是什么构成了那种天鹅绒般的柔滑、坚实的苹果、神秘的微笑,这些可能是未知的,或者还不够清楚,以至于无法发明一种算法来在图片中捕捉它们。事实上,有些人会说艺术、文学或音乐的创造力永远无法用算法来解释。

其次,算法揭示了必须遵循的每一个步骤。只有当算法的构成步骤在我们的意识中时,我们才能构建算法。但是,我们在骑自行车或抓住我们希望描绘的场景的细微差别时所执行的许多动作,都发生在认知科学家所说的"认知无意识"中。这种无意识行为能被提升到意识表面的程度是有限的。

再次,即使我们(合理地)很好地理解了任务的性质,该任务也可能涉及多个变量或参数,这些变量或参数以显要的方式相互影响。我们对这些相互作用的认识或理解可能是不完善、不完整,甚至是完全不足的。例如,设计计算机外部系统结构的问题(见第五章)就体现了这一特点。架构师可能很了解将进入外部系统结构的实际部分(数据类型、操作、内存系统、操作数寻址模式、指令格式、字长),但每个部分的变化范围以及这些部分对彼此的影响,可能只是不准确的或模糊的理解。事实上,理解这些交互的全部性质很可能会超出架构师的认知能力。

最后,即使一个人对问题有足够的理解,拥有关于问题域的知识,并且可以构造一种算法来解决问题,执行该算法所需的计算资源(时间或空间)也可能是完全不可行的。指数时间复杂度的算法(见第三章)就是例子。

下国际象棋就是一个很好的例子。它的问题性质很好理解,具有精确的合法行棋规则,并且是"完全信息"游戏,因为每个玩家都可以在每个时间点看到棋盘上的所有棋子。可能的结果是精确定义的:白赢,黑赢,或者平局。

但考虑到玩家的困境。每当轮到他上场时,他的理想目标是选择一个会获胜的走法。原则上,棋手可以遵循一个最优策略(算法):

> 此时轮到下棋的玩家自己考虑所有可能的走法。然后,对于每一步这样的走法,他都会考虑对手所有可能的走法;对于对手的每一步可能的走法,他都会再次考虑他所有

可能的走法；依此类推，直到达到最终状态：赢、输或平局。然后反推，玩家确定当前位置是否会获胜，并相应地选择一步走法，假设对手采取对自己最有利的走法。

这被称为"穷举搜索"或"暴力"方法。原则上这是可行的。当然，这是不切实际的。据估计，在典型的棋盘配置中，大约有30种可能的合法走法。假设一场典型的博弈持续了大约40步，然后其中一名棋手退出。我们从头开始考虑这个过程，玩家必须考虑30种可能的下一步走法；对于每一步，对手有30种可能的走法，即第二步有30^2种可能性；对于这30^2种选择中的每一种，第三步中还有另外30种选择，即30^3种可能性。依此类推，直到第40步，可能性的数量为30^{40}。所以一开始，玩家将不得不考虑$30 + 30^2 + 30^3 + 30^4 + ... + 30^{40}$种替代走法，然后再选择"最优"走法。替代路径的空间可以说是天文数字。

最终，如果要构建一个算法来解决问题，那么算法工作所需的有关问题的任何知识都必须完全嵌入算法中。正如我们在第三章中提到的，算法是一个自包含的程序知识。进行石蕊测试，执行纸笔乘法，计算数字的阶乘，从中缀算术表达式生成逆波兰式表达式（参见第四章）——所有需要知道的只是算法本身。如果不能将任何必要的知识纳入算法中，那么这个算法是不存在的。

这个世界充满了任务或问题，这些任务或问题体现了刚才提到的各种特性。它们不仅包括智力和创造性工作，比如科学研究、发明、设计、创意写作、数学工作、文学分析、历史研究；还包括专业从业者，比如医生、建筑师、工程师、工业设计师、规划师、教师、工匠所做的工作。即使是普通的、单调的活动，比如开车穿过繁忙的大道,对工作机会做出决定计划假期旅行,也不利于用算法解决,或者至少不利于用有效的算法解决。

然而,人们仍然在执行这些任务,解决这些问题。他们不等待算法,不管算法是否有效。事实上,如果我们必须等待算法来解决我们的所有问题,那么作为一个物种,我们早就灭绝了。从进化的角度来看,算法并不是我们思维方式的全部。所以问题出现了:我们可以使用哪些其他计算手段来执行这些任务?答案是求助于一种应用启发式的计算模式。

启发式是基于常识、经验、判断、类比、有根据的猜测等的规则、规范、原则和假设,它们提供了希望,但不能保证能解决问题。我们在上一章讨论计算机体系结构时遇到了启发式算法。然而,将计算机体系结构说成是一门基于启发式的人工科学是一回事,在自动计算中应用启发式算法是另一回事。我们现在考虑的是后者,也就是启发式计算。

搜索,可能会发现方案

启发式计算体现了一种冒险精神!启发式计算中存在不确定性和未知因素。一个寻找启发式解决方案的问题解决智能体(人或计算机)实际上处于一种未知领域之中。就像在未知物理领域中的某人进入探索或搜索模式一样,启发式智能体也是如此:智能体在计算机科学家所谓的问题空间中搜索问题的解决方案,但永远不确定能得到解决方案。因此,一种启发式计算也被称为启发式搜索。

例如,考虑以下的场景。你正在进入一个非常大的停车场,该停车场与您希望观看活动的礼堂相连。问题是找到一个停车位。汽车已经停在那里,但你显然不知道空位的分布或位置。那么应该怎么做呢?

在这种情况下,停车场实际上就是问题空间。而你所能做的就是,从字面上看,寻找一个空的地方。但是,与其漫无目的地随机搜索,你可能会决定采用"第一匹配"策略:进入你遇到的第一个可用空位。或者,你可以采用"最佳匹配"策略:找一个离礼堂最近的空位。

这些启发式方法有助于在问题空间中引导搜索。当然,这是有代价的。第一次匹配策略可能会减少搜索时间,但你可能需要走很长的路才能到达礼堂;最佳匹配策略可能需要更长的搜索时间,但如果成功,步行时间可能相对较短。当然,这两种启发式都不能保证成功。在这两种情况下,都可能找不到空位,在这种情况下,你可以无限期地搜索,也可以使用单独的标准终止搜索,例如"如果搜索时间超过限制则退出"。

然而,许多应用启发式的策略具有算法的所有特征(正如我们在第三章中讨论的那样),但二者有一个显著的区别:它们只对问题给出"近乎正确"的答案,或者,它们可能只对问题的某些实例给出正确的答案。因此,计算机科学家将某些启发式问题解决技术称为启发式或近似算法,在这种情况下,我们可能需要将它们与我们所谓的精确算法区分开来。术语"启发式计算"包括启发式搜索和启发式算法。稍后将介绍后者的一个实例。

一种称为"满足"的元启发式

通常,在优化问题中,目标是找到问题的最佳解决方案。许多优化问题都有精确的算法解决方案。不幸的是,这些算法通常具有指数时间的复杂度,因此用于问题的大型实例是不切实际的,甚至是不可行的。前面考虑的国际象棋问题就是一个例子。那么,如果最优算法在计算上不可行,那么问题求解者会做什么呢?

与其固执地追求最优目标,智能体可能会渴望实现更可行或更合理的目标,这些目标不是最优但是"可接受的好"。如果获得了满足这个愿望水平的解决方案,那么问题求解者就会感到满意。赫伯特·西蒙为这种心态创造了一个术语"满意度"。与优化相比,满意策略是一种更适度的抱负;它是选择可行的好而不是不可行的最优。

满意原则是一个非常高级、通用的启发式方法,可以作为识别更多特定领域启发式方法的跳板。我们可以称之为"元启发式"。

思索一种与国际象棋相关的满意度启发式算法。考虑到棋手的困境,正如我们所见,最优搜索被排除在外,需要更实用的策略,要求使用与国际象棋相关的(特定领域的)启发式原则。以下是其中最简单的。

考虑一个棋盘配置C(当前棋盘上所有棋子的位置)。使用一些"好的"度量G(C)来评估C的"承诺",G(C)考虑到C的一般特征(棋子的数量和种类、它们的相对位置等)。

假设M1,M2,...,Mn是配置C中棋手可以使出的走法,并假设每步走法后的结果配置是M1C、M2C等。然后选择一个使结果配置的价值最大化的走法,即选择优度值G(MiC)最高的Mi。

请注意,这里尝试了一种最优性。但这是一个"局部"或"短期"优化,只着眼于一步走法。这不是一个非常复杂的启发式,但它是一种休闲棋手可以培养的类型。但这确实需要玩家

（无论是人类还是计算机）对棋盘配置的相对优度有一定程度的深入了解。

国际象棋是满意度启发式搜索的例子。现在考虑一个令人满意的启发式算法的实例。

启发式算法

回想前一章对并行处理的讨论。只要满足伯恩斯坦的数据独立性条件，任务流中的一对任务 T_i、T_j（在任何抽象级别）都可以并行处理。

现在考虑一个由编译器从高级语言顺序程序（见第四章）为目标物理计算机生成的顺序机器指令流。但是，如果目标计算机可以同时执行指令，那么编译器还有一个任务要执行。识别指令流中指令之间的并行性并生成并行指令流，其中该指令流的每个元素都由一组可以并行执行的指令组成（称之为"并行集"）。

如果目标是最小化并行指令流中并行集的数量，这实际上是一个优化问题。像国际象棋问题一样，优化算法需要一个穷举搜索策略，因此在计算上是不切实际的。

在实践中，有许多令人更满意的启发式方法。一个例子就是我在这里所说的"先到先得"（FCFS）算法。

考虑以下顺序指令流 S。（为简单起见，此示例中没有迭代或 **goto** 语句。）

I1: A ← B;
I2: C ← D + E;
I3: B ← E + F − 1/W;

I4: Z ← C + Q;

I5: D ← A/X;

I6: R ← B − Q;

I7: S ← D∗Z.

FCFS算法如下。

FCFS:

  输入：一条直线顺序指令流S: <I1, I2, ..., In >；

  输出：一条直线并行指令流P，由一系列并行指令集组成，每一个指令集都存在于S中。

  对于S中以I1开始并以In结束的每条连续指令I，根据伯恩斯坦的数据独立性条件，将I置于最早可能存在的并行集合中。如果因为数据依赖性，排除了将I放入任何现有并行集的可能性，则在现有并行集之后创建一个新的（空）并行集并将I放置在那里。

  当将此FCFS算法应用于前面的示例时，可以看到输出是并行指令流P：

I1 ∥ I2;

I3 ∥ I4 ∥ I5;

I6 ∥ I7.

这里，"∥"表示并行处理，";"表示顺序处理。这是一个由三组并行指令组成的并行流。

  FCFS是一种令人满意的策略。它将每条指令放在尽可能早的并行集中，以便后续的数据相关指令也可以尽早出现在并行流中。令人满意的标准是："检查每条指令相对于其前指令的优点，并忽略后面的指令"。对于这个特定的例子，FCFS产生一个最优输出（并行集的最小序列）。但是，对于其他输入流，FCFS很可能会产生次优的并行集。

那么，精确算法和启发式算法有什么区别？在前一种情况下，"好"是通过评估其时间（或空间）复杂性来判断的。这些输出不存在意外或不确定性。对于同一任务（例如求解代数方程组、按升序对数据文件进行排序、处理工资单或计算两个整数的GCD等），两个或多个精确算法的输出不会发生改变。它们将（或可能）仅在性能和各自的审美吸引力上有所不同。然而，在启发式算法的情况下，还有更多的故事。当然，可以比较这些算法的相对时间复杂度。对于大小为n的输入流，FCFS是一种$O(n^2)$算法。但是，它们也可以根据输出进行比较，因为输出可能会引起意外。两个并行检测算法，或者两个使用不同启发式算法的国际象棋程序，可能会产生不同的结果。

它们的区别是一个经验问题。必须将算法实现为可执行程序，对各种测试数据进行实验，检查输出，并根据实验确定它们的优缺点。因此，启发式计算需要进行实验。

启发式和人工智能

人工智能是计算机科学的一个分支，涉及计算人工智能的理论、设计和实现，这些人工制品执行我们通常与人类思维相关的任务，这样的人工制品可以被视为"拥有"人工智能。因此，它在计算机科学和心理学之间架起了一座桥梁。对人工智能可能性的最早思考之一，是20世纪40年代末电气工程师克劳德·香农在考虑对计算机进行编程用以下国际象棋的想法时提出的。事实上，从那时起，计算机国际象棋一直是人工智能研究的重要焦点。然而，人工智能最具影响力的宣言（这个词本身是由该主题的先驱之一约翰·麦卡锡在20世纪50年代中期创造的）是1950年艾伦·图灵（以图灵机闻名）的一篇引发思考的文章，他抛出并提出了对这个问题的回答："声称计算机可以思考意味着什么？"他的回答涉及一种实验，即"思想实验"，在这个实验中，一个人通过一些"中性"通信手段（以便询问者无法从响应手段中猜测响应者的身份），其中一个智能体是人，另一个是计算机。如果询问者不能正确猜测计算机作为响应者的身份超过40%到50%，那么计算机可能被视为表现出类似人类的智能。这个测试后来被称为图灵测试，并且多年来一直是人工智能研究的圣杯。

人工智能是一个广阔的领域，事实上，人工智能研究人员青睐的范式不止一种。（我使用科学哲学家托马斯·库恩的"范式"一词。）然而，为了进一步阐明启发式计算的范围和能力，在这里将只考虑人工智能中的启发式搜索范式。

这种范式关注的是智能体——自然的和人工的，人类和机器。它基于范式的创始人艾伦·纽厄尔和赫伯特·西蒙最明确地阐述的两个假设：

物理符号系统假设：物理符号系统具有一般智能行为的必要和充分手段。

启发式搜索假设：物理符号系统通过对符号结构的问题空间进行渐进式和选择性（启发式）搜索来解决问题。

纽厄尔和西蒙所说的"物理符号系统",是指处理符号结构但又以物理基底为基础的系统,在这里我称之为物质和阈限计算人工制品,只不过它们包括自然的和人造的物体。

图8描绘了一个基于启发式搜索的问题解决智能体(人或人工的)的非常通用的结构描述。通过首先在工作区或问题空间中创建问题的符号表示来解决问题。问题表示通常标示初始状态,即智能体开始的位置,以及目标状态,表示问题的解决方案。此外,问题空间必须能够标示从初始状态到目标状态达到的所有可能状态。问题空间就是数学家所说的"状态空间"。

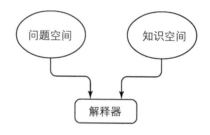

图8　启发式搜索系统的一般结构

从一种状态到另一种状态的转换,是通过诉诸智能体知识空间的内容(长期存储的内容)来实现的。选择该知识空间中的元素并将其应用于问题空间中的"当前状态",从而产生新状态。图8中显示了执行此操作的解释器(稍后会详细介绍)。知识元素的连续应用实际上导致了智能体在问题空间中进行求解,这个搜索过程构成了一个计算过程。当从初始状态开始,应用一系列知识元素导致达到目标状态时,问题就解决了。

然而,由于问题空间可能任意大,因此搜索不是随机进行的。相反,智能体应用启发式算法来控制搜索量,修剪掉不必要的搜索空间部分，从而尽快收敛到一个解决方案。

弱方法和强方法

因此，启发式搜索范式的核心是包含在知识空间中的启发式。这些范围可能从非常通用的（适用于广泛的问题域）到非常具体的（与特定的问题域相关）。前者被称为弱方法，后者被称为强方法。一般来说，当对问题域了解甚少时，弱方法更有希望；当问题域已知或被更详细地理解时，强方法更合适。

一种有效的弱方法（我们已经多次遇到过）是分而治之。另一种被称为手段-目的分析：

> 给定当前问题状态和目标状态，确定两者之间的差异。然后通过应用相关的"运算符"来减少差异。但是，如果不满足运算符应用的必要先决条件，则通过对当前状态和先决条件递归地应用手段-目的分析来减少当前状态和先决条件之间的差异。

一个可以同时应用分治和手段-目的分析的一个例子是学生计划她的学位课程。分治将问题分解为与学位课程的每一年相对应的子问题。最初的目标状态（例如，在X年内完成特定学科的毕业）被分解为X年中的每一年的"子目标"状态。对于每一年，学生将确定该年的初始状态（该年之前已经学习过的课程），并尝试确定要学习的课程，以消除该年初始状态和目标状态之间的差异。通过选择必修课程来缩小对课程的搜索范围。但其中的一些课程可能需要先决条件。因此，手段-目的分析被用来缩小初始状态和先决条件之间的差距,并依此类推。

请注意,手段-目的分析是一种递归策略(见第三章)。那么它有什么"启发式"呢?关键是不能保证在特定的问题域中,手段-目的分析能够成功终止。例如,给定当前状态和目标状态,可以应用几个操作来减少差异。选择的操作可能决定成功与失败之间的差异。

强方法通常代表问题域专家通过正规教育、实践培训和经验所拥有的专家知识。确定化学品的分子结构或辅助工程师设计项目的计算系统是典型的例子。这些启发式通常在知识空间中表示为以下形式的规则(称为产生式):

IF 条件 THEN 动作。

也就是说,如果问题空间中的当前状态与产生式的"条件"部分相匹配,则可以采取相应的"动作"。作为数字电路设计领域的示例(并作为启发式设计自动化系统的一部分实现):

IF　　电路模块的目标是将串行信号转换为并行信号
THEN　 使用移位寄存器。

当然,问题空间中的当前状态可能与多个产生式的条件部分匹配:

IF 条件 1 THEN 动作 1;
IF 条件 2 THEN 动作 2;

……

IF 条件 M THEN 动作 M。

在这种情况下,选择要采取的动作可能需要一个更高层级的启发式来指导(例如,选择第一个匹配的产生式)。正如稍后在计算中实现的那样,这可能是一个错误的选择,在这种情况下,系统必须"回溯"到先前的状态并探索其他一些产生式。

解释启发式规则

请注意图8中的"解释器"。它的任务是执行类似于物理计算机中的指令周期的循环算法(第五章):

匹配:识别知识空间中所有条件部分与问题空间中当前状态相匹配的产生式。将这些规则收集到一个冲突集中。

选择:根据选择启发式算法从冲突集中选择一个首选规则。

执行:执行首选规则的操作部分。

转到匹配步骤。

除了启发式搜索范式的不确定性之外,与算法(精确或启发式)的另一个显著区别是,(如前所述)在后者中,执行算法所需的所有知识都嵌入到算法本身中。相反,在启发式搜索范式中,几乎所有知识都位于知识空间(或长期存储器)中。启发式搜索范式的复杂性主要在于知识空间的丰富性。

第七章

计算思维

某种思维方式

现代——比如说第二次世界大战之后——的大多数科学都是技术性的,甚至是深奥的,以至于对它们更深入的理解仍然很大程度上局限于专家,即这些科学从业者的社群。例如,现代物理学中的基本粒子,充其量当与生活相关时,它们的影响才通过技术后果向广大公众揭示。

然而,有些科学则凭借其中心思想引人注目的性质,触动了专家之外的人们的想象力。进化论就是自然科学领域的一个例子。它的影响力已经延伸到社会学、心理学、经济学甚至计算机科学等与基因或自然选择无关的思想领域。

在人工科学中,计算机科学表现出相似的特征。我指的不是已殖民社会世界的无处不在的且"咄咄逼人"的技术工具。更确切地说,我指的是某种思维方式的出现。

人工智能的先驱之一西摩尔·派普特在他的著作《头脑风

暴》(1980)中热情而雄辩地阐述了这种思维方式,或者至少是它的前景。派普特宣称,他在这项工作中的目标是讨论和描述计算机如何为人类提供新的学习和思考方式,不仅作为一种实用的、工具性的人工制品,而是以一种更基本的、概念性的方式。即使思想家没有直接接触物理机器,这种影响也会促进思维模式。对派普特来说,计算机有望成为潜在的"强大思想的载体和文化变革的种子"。他认为,这本书将讲述计算机如何帮助人类有效地跨越客观知识与自我知识以及人文与科学之间的传统界限。

派普特所表达的是一种愿景,这种愿景也许是乌托邦式的,但它远远超出了计算机和计算对世界事务的纯粹工具性影响。后一种设想在19世纪中叶查尔斯·巴贝奇、阿达·洛芙莱斯伯爵夫人时代自动计算刚开始的时候就存在了。派普特的愿景,更确切地说,是灌输一种思维方式,这种思维方式将指导、塑造和影响一个人思考、感知和回应世界各个方面的方式,一个人的内心世界和外部世界,尽管表面上与计算没有明显的联系,但可能是通过类比、隐喻和想象的方式。

在派普特发表宣言二十五年后,计算机科学家周以真给这种思维方式起了一个名字:计算思维。但周以真的愿景可能比派普特的更平淡。她在2008年写道,计算思维涉及诸如解决问题、设计和理解利用计算基本概念的智能行为等活动的方法。然而,计算思维不能自成一座孤岛。在解决问题领域,它类似于数学思维;在设计领域,它将与工程思维方式共享特征;在理解智能系统(当然包括大脑)方面,它可能会与科学思维具有共同点。

与派普特一样,周以真将计算思维的思维方式与物理计算机本身分离开来,人们可以在没有计算机的情况下进行计算式思考。

但是,这种计算思维的思维方式意味着什么呢?我们稍后会看到一些例子,但在此之前,让我们跟随人工智能研究员保罗·罗森布鲁姆从两种关系的角度来解释计算思维的概念:一种是交互,这是之前介绍过的一个概念(见第二章)。用罗森布鲁姆的话来说,意思是两个实体之间的"相互动作、作用或影响"。然而,交互可以表示一个系统A对另一个系统B的单向影响(符号上,罗森布鲁姆将其描述为"A→B"或"B←A")以及双向或相互影响(符号"A←→B")。通过实现,罗森布鲁姆意在较高抽象层次上"实施"系统A,即在较低抽象层次上的系统B内的交互过程(符号上为"A/B")。实现的一个特殊情况是模拟:当B模仿或模拟A的行为时,表示为B模拟A(A/B)。

罗森布鲁姆解释说,使用这两个关系,计算思维的最简单表示是计算人工制品(C)影响人类行为(H):C→H。接着,罗森布鲁姆更进一步讨论。不只是一个人,假设我们考虑一个人模拟一个计算人工制品C:C/H。在这种情况下,我们有关系C→C/H,这意味着计算人工制品会影响模拟此类人工制品行为的人类。或者,我们可以更进一步:考虑一个人H在心理上模拟一个计算人工制品C,它本身已经实现或正在模拟某个现实世界域D的行为D:D/C/H。例如,假设D是人类行为。那么D/C意味着使用计算机来模仿或模拟人类行为。而D/C/H是指一个人在心理上模拟这种人类行为的计算机模型。这导致了对计算思维的以下解释:C→D/C/H。

可能有更细微的解释，但这些在交互和实现/模拟方面的解释足以说明计算思维的一般范围。

计算思维作为心理技能

计算机对人类最明显的影响，是作为一种心理技能的来源：一种分析和解决问题的工具，无论是否有实际的计算机，人类都可以在他们的生活过程中应用之。这就是周以真的想法。特别地，她将抽象视为计算思维的"本质"和"基本要素"。但是，虽然（正如我们在本书中看到的那样）抽象无疑是一个核心计算概念，但计算机科学提供了更多的概念，人类可以吸收并集成到自己的思维工具套件中。我在想启发式方法，有弱方法和强方法；将满意而不是优化作为现实决策目标的想法；以算法的方式思考并理解何时以及是否是解决问题的适当途径；并行处理作为实现多任务处理的条件和架构；"自上而下"（从目标和初始问题状态开始，并将目标细化为更简单的子目标，再将后者进一步简化为更简单的子目标等）或"自下而上"（从目标和最低层的构建块开始，通过将构建块组合成更大的构建块来构建解决方案等）来处理问题。但重要的是，要获得这些思维工具，需要对计算机科学概念有一定程度的掌握。周以真认为，这需要从小就引入计算思维作为教育课程的一部分。

但计算思维不仅仅需要分析和解决问题的能力。它包含一种想象的方式，通过看到类比和构建隐喻的方式。我认为，派普特想到的正是这种技术技能和想象力的结合，它提供了计算思维的充分丰富的思维方式。我们现在考虑一些智力和科学探究领域，在这些领域中，这种思维方式已被证明是有效的。

关于思维的计算式思考

当然,尽管存在争议,但这种思维方式最有力的表现之一是对思考的思考:计算机科学对认知心理学的影响。将图灵的著名问题,即计算机是否可以思考(人工智能的基础)反过来,认知心理学家思考了这个问题:思考是一个计算过程吗?

对这个问题的回答,可以追溯到20世纪50年代末艾伦·纽厄尔和赫伯特·西蒙的开创性工作,他们开发了一种人类解决问题的信息处理理论,将启发式、抽象层次和符号结构等计算问题与逻辑结合起来。最近,它促进了认知架构模型的构建,其中最突出的是心理学家约翰·安德森和计算机科学家艾伦·纽厄尔、约翰·莱尔德和保罗·罗森布鲁姆等研究人员。安德森的模型系列一般被称为ACT,纽厄尔等人的模型被称为SOAR,这些都受到计算机内部系统结构基本原则的强烈影响(见第五章)。在这些模型中,认知的架构被探索为持有代表世界各个方面符号结构的存储层级结构,以及通过类似于指令周期的过程对符号结构进行操作和处理。这些架构模型已经在理论上和经验上被广泛研究,在一定抽象层次上作为思维的可能理论。

另一种受计算影响的思维模型始于并行处理和分布式计算的原理,并将思维想象为一个由分布式、交流和交互的认知模块组成的"社会"。这种心理模型的一个有影响力的支持者是人工智能先驱马文·明斯基。认知科学家和哲学家玛格丽特·博登则将其在认知科学史上的权威著作命名为《思维即机器》(2006):根据她的描述,思维是一种计算设备。

计算大脑

将大脑的神经元结构表示或建模为一个计算系统，相反地，将计算人工制品表示为高度抽象的神经元样实体的网络，其历史可以追溯至20世纪40年代的数学家沃伦·皮茨和神经生理学家沃伦·麦卡洛克，以及非常著名的约翰·冯·诺伊曼的开创性工作。在接下来的六十年里，一种被称为连接主义的科学范式得以发展起来。在这种方法中，计算思维的思维方式最具体地体现在设计高度互连的网络中（因此称之为"连接主义"），这些网络由非常简单的计算元素组成，这些元素共同用于模拟基本大脑过程的行为，这些基本大脑过程是高级认知过程（例如，检测线索或识别视觉过程中的模式）的构建模块。与前一节提到的符号处理认知结构相比，大脑的连接主义架构处于较低的抽象层次。

认知科学的出现

头脑的符号处理认知架构和大脑的连接主义模型，是计算人工制品和计算机科学原理影响认知科学，这一相对较新的跨学科领域的形成和出现的两种方式。我必须强调，并非所有认知科学家，例如心理学家杰罗姆·布鲁纳，都将计算视为认知的核心要素。尽管如此，通过构建计算模型和基于计算的假设来理解诸如思考、记忆、计划、解决问题、决策、感知、概念化和理解等活动的想法是令人信服的。特别是，将计算机科学视为一门自动符号处理科学，这一观点在认知科学本身的出现起到了强大的催化剂作用。前一节提到的玛格丽特·博登，认为认知科学史的核心是自动计算的发展。

了解人类的创造力

创造力这个迷人的主题，从特殊的、历史上原创的类型，到个人的日常品牌，是一个广泛的话题，吸引了心理学家、精神分析家、哲学家、教育家、美学家、艺术理论家、设计理论家、知识史学家和传记作家的专业关注；更不用说那些自我反思的创造者本身（科学家、发明家、诗人、作家、音乐家、艺术家等）。因此，创造力的方法、模型和理论的范围之广令人眼花缭乱，尤其是对创造力的定义多种多样。

但至少有一个创造性研究人员社群将计算思维作为一种操作方式。他们提出了创造性过程的计算模型和理论，这些模型和理论很大程度上借鉴了启发式计算的原则，将知识表示为复杂的符号结构（称为模式）和抽象原理。在这里，由于其引人注目的影响，计算思维为科学、技术、艺术、文学和音乐创造力的分析提供了一个共同基础，正如派普特所希望的那样，多种文化的结合。

例如，文学学者马克·特纳将计算原理应用于理解文学作品的问题，正如科学哲学家和认知科学家保罗·萨加德努力通过计算模型来解释科学革命，本文作者是计算机科学家和创造力研究者，为人工科学中的技术人工制品和思想的设计和发明构建了一个计算解释。计算思维的思维方式已成为将这些不同的知识和创造性文化结合为一体的黏合剂。在许多关于创造力的计算研究中，计算机科学提供了一种精确的思维方式，可以表达与创造力有关的概念，而这在以前是不存在的。

举个例子，作家阿瑟·库斯勒在他的巨著《创造的艺术》（1964）中假设了一个被称为"混联"的过程，作为实现创造性艺术的机制。库斯勒所说的"混联"指的是将两个或多个不相关的概念结合在一起，并将其混合，从而产生一个原创产品。然而，究竟是如何发生混联仍然无法解释。计算思维为一些创造性研究人员（如马克·特纳和本文作者）提供了用计算机科学的精确语言解释某些混联的方法。

了解分子信息处理

众所周知，1953年，詹姆斯·沃森和弗朗西斯·克里克发现了脱氧核糖核酸（DNA）分子的结构。分子生物学的科学由此诞生。它的关注点包括了解和发现诸如DNA复制、DNA转录为RNA以及RNA翻译为蛋白质基础生物过程等机制。因此，分子作为信息载体的概念进入了生物意识。受计算思想影响的理论生物学家，开始用计算术语对遗传过程进行建模（顺便说一句，这也促进了基于遗传概念的算法的发明）。计算思维塑造了所谓的"生物信息处理"，或者用当代术语来说，就是生物信息学。

结语　计算机科学是一门通用科学吗？

贯穿本书的前提是计算机科学是一门人工科学。它以符号处理（或计算）人工制品为中心；它是一门关于事物应该如何而不是事物实际如何的科学；在理解这门科学的本质时，必须考虑到设计者（算法设计师、程序员、软件工程师、计算机架构师、信息学家）的目标。在所有这些方面，与自然科学的区别是显而易见的。

然而，在前一章中，我们已经看到计算思维充当了计算人工制品世界与自然世界之间的桥梁，特别是生物分子、人类认知和神经元过程之间的桥梁。那么，问题是，计算不仅提供了一种思维方式，而在潜移默化中，是否计算作为一种现象，包含了自然和人工科学呢？也就是说，计算机科学是一门通用科学吗？

近年来，一些计算机科学家正是沿着这些思路思考的。因此，彼得·丹宁认为，计算不应再被认为是一门人工科学，因为信息处理在自然界中比比皆是。丹宁和另一位计算机科学家彼得·弗里曼认为，在过去的几十年里，（一些计算机科学家的）

注意力的焦点已经从计算人工制品转移到信息过程本身——包括自然过程。

对于丹宁、弗里曼和另一位计算机科学家理查德·斯诺德格拉斯而言，计算是一门自然科学，因为计算机科学家不仅要阐明事物应该是怎样的，而且还要发现事物是怎样的（在大脑中，在活细胞中，甚至在计算人工制品领域）。这种观点意味着计算人工制品与自然实体属于同一本体论类别。或者说，自然和人造之间没有区别。事实上，斯诺德格拉斯发明了一个词来描述计算机科学属于自然科学，即"Ergalics"，它来自希腊语词根"ergon"（εργων），意思是"工作"。

保罗·罗森布鲁姆大体上同意斯诺德格拉斯的观点，但为了避免出现新词，他简单地将计算机科学与物理、生命和社会科学列为"第四大科学领域"。

计算机科学作为构成其自身范式的独特性一直是本书的永恒主题，因此罗森布鲁姆的论文与这个主题是一致的。问题是，是否应该区分对自然信息过程的研究和对人工符号过程的研究。在这里，信息和符号之间的区别似乎是合理的。在自然域中，实体只代表它们自己。诸如神经元之类的实体，或作为DNA组成部分的核苷酸，或构成蛋白质的氨基酸，除了它们本身之外不代表任何东西。因此，我发现将DNA处理称为符号处理是有问题的，尽管将这些实体称为非参考信息的载体似乎是正确的。

我认为，在本体论上，必须区分作为人工科学的计算机科学和作为自然科学的计算机科学。在前者中，人类能动性（以目标和目的、获取知识、影响行动的形式）是科学的一部分。在后一种情况下，智能体显然不存在。范式从根本上是不同的。

尽管如此，且不管存在任何这种可能的本体论差异，计算机科学带给我们的东西，正如前几章试图展示的那样，是一种非常独特的感知、思考和解决广泛问题的方式——跨越了自然、社会、文化、技术和经济领域。这无疑是它对现代世界最原始的科学贡献。

索 引

(条目后的数字为原书页码，
见本书边码)

A

abstract artefact 抽象人工制品 23—25,
28, 31, 41—42, 53, 62, 68, 82

abstract expressionism 抽象表现主义
104

abstract time 抽象时间 53, 55

abstraction 抽象 18—20, 22, 39, 41, 59, 62,
81, 85, 123—124, 126

accountancy 会计 20

Ackoff, Russell 罗素·阿科夫 8

Act of Creation, The《创造的艺术》126

addressing modes 寻址模式 88, 105

aesthetics 美学 57—59, 74—75, 112

agent 智能体 29, 107, 114

Aho, Alfred 阿尔弗雷德·阿霍 56

algorism 循序渐进的过程 34

algorithm 算法 24, 33, 34—59, 62, 64, 75,
83, 94, 104—113, 118

algorithmic thinking 算法思维 34, 62,
104

al-Khwarizmi, Mohammed ibn-
Musa 阿布·阿卜杜拉·穆罕默
德·伊本·穆萨·花拉子密 33

Amdahl, Gene 吉恩·阿姆达尔 84, 100

Amdahl's law 阿姆达尔定律 100

amino acids 胺基酸 130

analogical reasoning 类比推理 46

analogy 类比 63—64, 82, 120

Analytical Engine 分析机 2

approximate algorithm 近似算法 108

architecture 体系结构 15, 17

arithmetic expression 算术表达式
48—49, 73, 106

arithmetic instructions 算术指令 87

arithmetic operations 算术运算 50—51,
71, 91—92

array data structure 数组数据结构 71

Art of Computer Programming, The《计
算机程序设计艺术》74

artefact 人 工 制 品 2, 5, 13—14, 17—19,
21—25, 28, 31—32, 41—42, 46, 53, 63, 68—69,
78, 81, 95, 129

artificer 技师 19, 95, 129

artificial intelligence（AI）人工智能
2, 6, 11, 64, 113—117, 120

artificial language 人造语言 64, 67

artificial process 人工处理 130

assembler 汇编程序 68

assembly language 汇编语言 68, 85, 89

assignment 赋值 70—71, 73

assignment operation 赋值操作 36

assignment statement 赋值语句 70,
72—73

astronomy 天文学 33

automata theory 自动机理论 29—30

automatic computation 自动计算 xv,
10, 64, 120, 125

automaton（automata）机器人 1—3,
12—13, 24

average performance 平均性能 54

axiomatic tradition 公理传统 77, 80

axioms 公理 43, 63, 76—77

125

axioms of arithmetic 算术公理 43

B

Babbage, Charles 查尔斯·巴贝奇 xv, 2, 120
backtracking 反推 117
beauty 美 75, 77
Bernstein, A.J. 伯恩斯坦 97
Bernstein's conditions 伯恩斯坦条件 97
best fit policy 最佳匹配节点策略 108
big data 大数据 11
Big O notation 大 O 表示法 55—56, 113
binary digit (bit) 二进制位 4, 12, 87
binary search 二分搜索 56—57, 59
bioengineering 生物工程学 31
bioinformatics 生物信息学 127
biology 生物学 64
bit string 位串 91
Blaauw, Gerrit 格里特·布洛乌 84
Boden, Margaret 玛格丽特·博登 124—125
Böhm, Corrado 科拉多·伯姆 74—75
Boolean expressions 布尔表达式 61
bottom up approach 自下而上的方法 122
brain 大脑 64, 124—125
branch instruction 分支指令 87
Brooks, Frederick 弗雷德里克·布鲁克斯 84
Bruner, Jerome 杰罗姆·布鲁纳 125
brute force search 穷举搜索 106
byte 字节 4, 12, 87

C

cache memory 高速缓冲存储器 92
character string 字符串 71—72, 92
chemical notation 化学符号 67
chemistry 化学 3
chess 国际象棋 105, 109—110, 113
Chomsky, Noam 诺姆·乔姆斯基 75
Church, Alonzo 阿隆佐·丘奇 xv, 28
Church-Turing thesis 丘奇-图灵论题 28
circuit design 电路设计 95, 117
circuit theory 电路理论 31
civil engineering 土木工程 3, 20
cognition 认知 64, 125, 129
cognitive architecture 认知架构 123—125
cognitive process 认知过程 63
cognitive psychology 认知心理学 64, 123
cognitive science 认知科学 xvi, 64, 125
cognitive scientist 认知科学家 63
cognitive unconscious 认知的无意识 104
communication 通信 65
communication network 通信网络 25, 91
compiler 编译器 47—49, 65, 69, 76, 85, 102
complexity 复杂度 14, 54—55, 59—61, 105, 112—113
compositional hierarchy 组成层级 14—18, 22, 81
computation 计算 xvii, 10, 62, 64—65, 68, 72, 115, 125

computational artefact 计算人工制品 xvii, 5, 13—14, 17—19, 21—23, 29—32, 42, 64, 78, 81, 83, 114, 121—122, 124—125, 129—130

computational complexity 计算复杂度 59—61

computational concepts 计算的概念 69

computational language 计算语言 67, 69

computational problems 计算问题 59—61

computational speedup 计算速度 96

computational thinking 计算思维 34, 119—127, 129

computational throughput 计算吞吐量 96

computer, physical 物理计算机 xv, xvi, xvii, 1—5, 13, 20—21, 24, 47, 49, 62, 64—65, 69, 76, 81, 83—85, 89, 95, 120

computer architect 计算机架构师 83, 90, 129

computer architecture 计算机体系结构 15, 24, 65, 81—103, 105, 124

computer design 计算机设计 65

computer hardware description language 计算机硬件描述语言 69

computer design and description language 计算机设计与描述语言 69

computer network 计算机网络 5, 84

computer program 计算机程序 18, 22—23, 47, 62, 64, 68, 76

computer programming 计算机编程 9, 62—79

computer science 计算机科学 xv, xvi, xvii, 1—3, 6, 8—9, 11—12, 14, 19, 23, 30, 32, 34, 59, 62, 64—65, 67, 83, 113, 119, 123, 125—126, 129

computer scientist 计算机科学家 xvi, 5, 8—11, 14, 29, 34, 36, 44, 58, 62, 67, 73, 81, 130

computer system 计算机系统 14, 78

computing 计算 xv, xvi, 2—3, 9, 12—13, 15, 70

concepts and categories 概念和范畴 69—74

conditional statement 条件语句 74, 96

connectionism 连接主义 124—125

constructive hierarchy 构造层级 19—22, 81

control unit 控制单元 91, 94—95

Cook, Stephen 斯蒂芬·库克 60—61

cores 核 99

correctness 正确性 53, 77, 80

creativity 创造力 46, 104, 125—126

Crick, Francis 弗朗西斯·克里克 126

D

Darwinian theory of evolution 达尔文进化论 34, 64

data 数据 8—13, 21—22, 30, 72, 92

data dependency relations 数据依赖关系 97—98, 100

data format 数据格式 88

data mining 数据挖掘 6—8, 11

data object 数据对象 10—11, 71, 88, 92, 96

data processing 数据处理 11

data stream 数据流 98
data structure 数据结构 10—11, 72
data type 数据类型 10—11, 22, 71—72, 86—88, 105
database 数据库 7, 10—11, 13
decision procedure 决策过程 34
declarative knowledge 陈述性知识 11, 43—45, 47
deductive logic 演绎逻辑 76
definiteness (of algorithms)（算法的）确定性 36, 38, 41
definitions 定义 43, 63, 77
Denning, Peter 彼得·丹宁 129—130
design 设计 42, 45—47, 50, 53, 58, 64—65, 74, 79, 82, 102, 121
Design and Analysis of Computer Algorithms, The《计算机算法的设计与分析》56
determinacy (of algorithms)（算法的）确定性 41—42
Dijkstra, Edsger 艾兹格·迪杰斯特拉 19, 58, 75—76
distributed architecture 分布式架构 25
distributed computing 分布式计算 84, 124
divide-and-rule 分而治之 50, 116
digital circuit 数字电路 89
DNA 脱氧核糖核酸 126—127, 130

E

economics 经济学 77, 119
effectiveness (of algorithms)（算法的）有效性 36, 38
efficiency (of algorithms)（算法的）效率 55
electronic mail (email) 电子邮件 15, 17, 20, 83
electronics 电子 95
Elements, The (Euclid)《几何原本》（欧几里得）33
Eliot, T.S. 艾略特 6—7
empirical object 实证对象 63
empirical science 实证科学 xvi
engineering 工程 30, 63, 80, 121
engineering drawings 工程图纸 46
engineering science 工程科学 31, 80
environment 环境 83
Ergalics 工作 130
Ershov, A.P. 叶尔绍夫 75
Euclid 欧几里得 33
Euclid's GCD algorithm 欧几里得的GCD算法 36—38, 73—74
evidence 证据 47
exact algorithm 精确算法 109, 112
exhaustive search 穷举搜索 106, 111
experiment 实验 8, 32, 102, 113
expert knowledge 专家知识 116
exponential function 指数函数 56, 59, 105
expression 表达式 48—49, 51, 73

F

fabrication technology 制造技术 95
factorial 阶乘 44—45, 58
facts 事实 43

file 文件 17
finiteness (of algorithms)（算法的）有限性 36—38, 40
first come first served algorithm 先到先得算法 111
first fit policy 第一匹配策略 108
Floridi, Luciano 卢恰诺·弗洛里迪 7—9
Freeman, Peter 彼得·弗里曼 129—130
function 函数 51
functional style of programming 函数式编程 25
fundamental particles 基本粒子 12

G

genetic algorithm 遗传算法 127
genetic engineering 基因工程 31
gigabyte 千兆字节 5
goto statement goto 语句 37, 74, 96
goal 目标 32
goal state 目标状态 114, 116
greatest common divisor (GCD) 最大公约数 33

H

hardware 硬件 4, 22, 25, 31, 81, 82
hardware description language 硬件描述语言 24
hardwire 硬连线 20
Hardy, G.H. 哈代 57
Hartmanis, Juris 尤里斯·哈特马尼斯 60

Hero of Alexander 亚历山大英雄 2
heuristic algorithm 启发式算法 108, 110—113
heuristic computing 启发式计算 104, 107—118, 126
heuristic search 启发式搜索 107—110, 114—118, 122
heuristic thinking 启发式思维 47, 101—102
heuristics 启发式 101, 107, 115, 123
hierarchical levels 层级 14, 81, 85
hierarchical organization 层级组织 14, 90
hierarchy 层级制度 6, 14, 18—19, 22, 81
Higgs boson 希格斯玻色子 7, 12
high level language 高级语言 47, 68, 69
Hindu art of reckoning 印度的计算艺术 33
Hoare, C.A.R. 霍尔 58, 76
Hoare's manifesto 霍尔的宣言 76—77
homunculus 人造人 93
Hopcroft, John 约翰·霍普克罗夫特 56
House of Wisdom, Baghdad 巴格达的智慧之家 33
human-computer interface 人机界面 29
human needs and goals 人的需要和目标 31
human thinking 人类思维 2
human understanding 人类理解 10
humanities 人文学科 120
hypothesis 假说 114

I

ignorance 无知 41
implementation 实施 18, 22, 40, 45, 52, 62, 65, 79, 85, 121—122
Industrial Revolution 工业革命 2
inference 推断 6
infix expression 中缀表达式 48—49, 51, 106
informaticist 信息学家 129
Informatik 信息学 xv, 3
information 信息 3—13, 21, 30, 40, 72, 88, 105
information, meaningless 无意义信息 4—5
information, semantic 语义信息 5
information age 信息时代 xvi
information organization 信息组织 11
information processes 信息处理 xv, 3—6, 9, 11, 123, 126, 129—130
information retrieval 信息检索 7, 11
information revolution 信息革命 xvi
information science 信息科学 11
information society 信息社会 xvi
information structure 信息结构 11
information system 信息系统 11
information technology 信息技术 xvi
information theory 信息理论 4, 11—12
informatique 计算机科学 xv, 3
inner (computer) architecture 内部（计算机）系统结构 85—86, 89—95, 124
input/output 输入/输出 17—18, 25, 36, 91
instruction cycle 指令周期 93—94, 98—99, 124
instruction execution unit 指令执行单元 91, 93
instruction interpretation unit 指令解释单元 91—94
instruction format 指令格式 88, 105
instruction pipeline 指令流水线 99
instruction register 指令寄存器 93
instruction set 指令集 86, 88
instruction stream 指令流 98—99, 110
instructions 指令 22, 87, 92
integer data type 整数数据类型 71, 87
integrated circuit 集成电路 95
interactive computing 交互计算 29—30, 121—122
interconnection network 互连网络 100
International Federation for Information Processing (IFIP) 国际信息处理联合会 3
Internet 互联网 5, 15, 83, 84
interpreter 解释器 15, 17—18, 115, 118
intractable problems 棘手问题 59—61
invention 发明 42
iteration 迭代 37, 45, 55, 96
iteration statement 迭代语句 74

J

Jacopini, Giuseppe 朱塞佩·亚科皮尼 74—75
Janus 雅努斯（雅努斯是罗马的门神，他看起来同时在两个相反的方向）42, 63

K

Karp, Richard 理查德·卡普 60
Keats, John 约翰·济慈 57
knowledge 知识 5—7, 10—11, 13, 21, 30, 42—45, 47, 105—106, 120
knowledge base 知识库 11
knowledge discovery 知识发现 6.7
knowledge engineering 知识工程 11
knowledge level 知识水平 11
knowledge processing 知识处理 6—7
knowledge representation 知识表示 11, 126
knowledge space 知识空间 46, 48—49, 101, 115, 118
knowledge structure 知识结构 11
knowledge system 知识体系 11
Knuth, Donald E. 高德纳 8—10, 12, 34, 58, 74—75
Koestler, Arthur 阿瑟·库斯勒 126

L

Laird, John 约翰·莱尔德 123
Langley, Patrick 帕特·兰利 9
language 语言 24, 38, 47, 64—65, 67—74
language design 语言设计 65
laptop 笔记本电脑 90
law 法律 43
laws of engineering 工程学定律 42
laws of motion 运动定律 22
laws of nature 自然法则 31
laws of physics and chemistry 物理和化学定律 23, 42, 81

laws of thermodynamics 热力学定律 22
learning 学习 120
levels of parallelism 并行级别 98
liminal computational artefact 阈限计算人工制品 23—24, 31, 62—64, 78, 83, 87, 101, 114
linear search 线性搜索 54, 57, 59, 71
linguistic categories 语言类别 69
linguistic relativity 语言相对论 67
linguistics 语言学 65
literate programming 文学编程 75
litmus test 石蕊测试 34—35, 41—42, 106
local area network 局域网 84
logic 逻辑 30, 47, 76, 123
logic circuit 逻辑电路 25
logic level 逻辑层 90
logic of design 设计逻辑 46
logical expressions 逻辑表达式 61
long-term memory 长期存储器 17—18, 21, 87—88, 115, 118
Lovelace, Ada Augustus 奥古斯塔·阿达·洛芙莱斯 2. 120
low-level language 低级语言 68
Lukasiewiz, Jan 扬·卢卡西维茨 49

M

McCarthy, John 约翰·麦卡锡 113
McCulloch, Warren 沃伦·麦卡洛克 124
Machina Speculatrix 机器冒险者 2
machine code 机器码 47, 49, 102
machine-dependent language 机器相

关语言 68
machine instructions 机器指令 22, 89, 110
machine-independent language 机器无关语言 68—69
magical realism 魔幻现实主义 104
material computational artefact 材料计算人工制品 22, 24—25, 31, 46, 81, 114
mathematical activity 数学活动 63, 121
mathematical expression 数学表达式 76
mathematical induction 数学归纳法 43
mathematical logic 数理逻辑 30
mathematical machine 数学机器 76
mathematical notation 数学符号 67
mathematical objects 数学对象 63
mathematical theory 数学理论 76
mathematics 数学 30, 33, 58
means and ends 手段和目的 32
means-ends-analysis 手段-目的分析 116
mechanical engineering 机械工程 20, 80
medium-term (main) memory 中期（主）存储器 87—89, 92, 96
megabit 兆位 5
memory 存储器 17, 21—22, 87, 89, 92
memory address 存储地址 87
memory hierarchy 存储层级 90, 124
memory management 存储管理 21
memory system 存储系统 90, 96, 100, 105
mentality 思维方式 68, 119—121, 123

meta-heuristic 元启发式 109
metalanguage 元语言 24
metaphor 比喻 35, 120, 123
methodology 方法 24
microarchitecture 微架构 90
microprogram 微程序 95, 99
microprogrammed control unit 微程序控制单元 95
microprogrammer 微程序设计员 95
microprogramming 微编程 94—95
microprogramming language 微程序设计语言 24
mind 思维 63—64, 124—125
Mind as Machine《思维即机器》124
mindless thinking 盲目的思考 42
Mindstorms《头脑风暴》120
Minsky, Marvin 马文·明斯基 124
modules 模块 79, 87, 92, 100
molecular biology 分子生物学 xvi, 127
molecular information processing 分子信息处理 126—127
Molière 莫里哀 33
Moore, Gordon 戈登·摩尔 95
Moore's law 摩尔定律 95
multicomputing 多计算 4
multiplication 乘法 39—40, 106
multiprocessor architecture 多处理器架构 24, 100

N

natural language 自然语言 64—65, 67, 69, 75
natural object 自然对象 63

natural process 自然过程 130
natural science 自然科学 30—31, 119, 129, 131
natural selection 自然选择 119
need 需要 32
network protocols 网络协议 84
neuclotide 核苷酸 130
neuron 神经元 124, 130
neuroscientist 神经科学家 64
Newcomen, Thomas 托马斯·纽科门 2
Newell, Allen 艾伦·纽厄尔 1, 12, 15, 102, 114, 123
non-deterministic polynomial time 非确定性多项式时间 60
non-recursive algorithm 非递归算法 59
notation 符号 36, 38, 49, 67, 69, 74
NP class of problems NP 类问题 60—61
NP-completeness NP 完备 60—61

O

objective knowledge 客观知识 42, 45, 120
observation 观察 9, 32
Ode on a Grecian Urn 《希腊古瓮颂》 57
Ohm's law 欧姆定律 22
ontology 本体 130—131
operand 操作数 50—51, 93, 105
operating system 操作系统 21—23, 78, 85
operator 运算符 50
optimization 优化 110—111, 122

outer (computer) architecture 外部（计算机）系统结构 84—89, 105

P

P class of problems P 类问题 60
P = NP problem, the P = NP 问题 61
palindrome 回文 28
Papert, Seymour 西摩尔·派普特 120—121, 123, 126
paradigm 范式 33, 114—115, 118, 131
parallel computing 并行计算 95—100
parallel processing 并行处理 96, 98—100, 110, 112, 122, 124
parallel program 并行程序 98
parenthesis-free expression 无括号表达式 48
parity detection 奇偶校验检测 26
Peano, Guiseppe 朱塞佩·皮亚诺 43
perfect information 完全信息 105
performance 性能 53
Perlis, Alan 艾伦·佩利 1, 12, 15
philosopher 哲学家 63
physical metallurgy 物理冶金学 31
physical symbol system hypothesis 物理符号系统假设 114
physical time 物理时间 53
physics 物理 3, 12, 77, 95, 119
pipeline processing 流水线处理 99
pipelined architecture 流水线体系结构 99
Pitts, Warren 沃伦·皮茨 124
pocket calculator 袖珍计算器 40—42
Polish notation 波兰式表示法 49—50

Polya, George 乔治·波利亚 101
polynomial time 多项式时间 59, 61
polynomial time problems 多项式时间问题 59—60
Popper, Karl 卡尔·波普尔 42, 103
precedence rules 优先规则 48—49
primitive data type 原始数据类型 71
private memory 私有存储 18
problem classes 问题类 60
problem domain 问题域 105
problem size 问题规模 53, 55
problem solving 问题解决 50, 107—108, 120—123
problem space 问题空间 107, 114—115
procedural knowledge 程序知识 11, 42—45, 47, 106
procedure 程序 35
process 处理 62
processing unit 处理单元 92—93, 96
production rules 产生规则 117
program behaviour 程序行为 92
program correctness 程序正确性 52—53, 77, 80
program counter 程序计数器 93
programmer 程序员 9—11, 19, 83, 129
programming as art 编程艺术 74—75
programming as engineering 编程作为工程 77—80
programming as mathematical science 编程作为数学科学 74—77
programming language 编程语言 24, 47, 64—65, 67—74, 98
programming theorist 编程理论家 9, 11

proof 证明 52, 57, 63, 76—77, 80
psychology 心理学 119
Pythagoras's theorem 毕达哥拉斯定理 44

R

read/write head 读写头 25
real number 实数 92
reasoning 推理 47
recipe 配方 41
recursion 递归 50, 52, 58
recursive algorithm 递归算法 51, 58—59, 116
refinement 细化 18, 81
registers 寄存器 87, 89, 92, 96
reliability 可靠性 78
representation 表示 8
requirements (for design) 需求（设计） 45, 79
reverse Polish 逆波兰式 49, 51
RNA 核糖核酸 127
Rock, The 《岩石》 6
Rosenbloom, Paul 保罗·罗森布鲁姆 5, 11, 121, 123, 130
rules 规则 10, 47—48, 73
rules of composition 组合规则 73—74
rules of semantics 语义规则 76
rules of syntax 语法规则 75

S

Sapir, Edward 爱德华·萨皮尔 67
Sapir-Whorf hypothesis 萨皮尔-沃尔

夫假说 67
satisfiability problem, the 满意度问题 60
satisficing 满意 109—112, 122
schema 模式 126
science 科学 xvii, 8, 30, 47, 120, 129—130
science of design 设计科学 46—47
science of programming 编程科学 75—77
sciences of the artificial 人工科学 30—32, 64, 101, 119, 129—130
scientific problem solving 科学的问题解决 103
scientific revolutions 科学革命 126
search 搜索 54, 56—57, 59, 106—108, 115, 118
self-knowledge 自我知识 120
semantic gap 语义鸿沟 90
semantics 语义 9, 76, 79
sequential machine 时序机 74
sequential processing 顺序处理 112
sequential statement 顺序语句 73—74
sequential task stream 顺序任务流 96
Shannon, Claude 克劳德·香农 4
shared memory 共享存储 18, 100
short-term-memory 短期存储器 88, 92
Shrager, Jeffrey 杰弗里·施拉格 9
Simon, Herbert 赫伯特·西蒙 1, 12, 14—15, 31, 102, 109, 114, 123
simulation 模拟 103, 121—122
smart phone 智能手机 40, 84
Snodgrass, Richard 理查德·斯诺德格拉斯 130
social media 社交媒体 83

sociology 社会学 119
software 软件 18—20, 22, 24—25, 78, 80—81
software developer/engineer 软件开发人员/工程师 19—20, 83, 129
software engineering 软件工程 78—80
software engineering environment 软件工程环境 79
software life cycle 软件生命周期 78—79
software maintenance 软件维护 79
software modification 软件修改 79
software specification 软件规范 79
software system 软件系统 21, 79
software verification and validation 软件的验证和确认 79
solipsistic computers 唯我论的计算机 83—84
sociable computers 社交计算机 83—84
space complexity 空间复杂度 54, 59, 112
speedup 加速 96, 100
state space 状态空间 115
state table 状态表 25—26
statement type 状态类型 73—74
statements 状态 72
steam engine 蒸汽机 2
Stearns, Richard 理查德·斯特恩斯 60
strength of materials 材料强度 31
strong method 强方法 115—117, 122
structural engineering 结构工程 80
structured data type 结构化数据类型 71
style 风格 74
subproblem 子问题 51

supercomputer 超级计算机 90
symbol processing 符号处理 12—13, 28, 70, 81, 124—125, 129—130
symbol processor 符号处理器 17, 81
symbol structure 符号结构 11—13, 23, 30, 41, 45—46, 68—69, 123—124, 126
symbol system 符号系统 114
symbols 符号 11—12, 25—26, 36, 69
syntax 语法 9, 75—76, 79
system program 系统程序 78

T

technology 技术 82, 86, 95—96, 126
termination (of algorithm)（算法的）终止 38, 45
Thagard, Paul 保罗·萨加德 126
theorems 定理 43, 63, 76—77
theory 理论 45, 47
theory of evolution 进化论 119
theory of knowledge 知识论 11
theory of structures 结构理论 31
theory of syntax 语法理论 75—76
thermodynamics 热力学 31
thinking machine 思考的机器 2
thought experiment 思想实验 113
throughput 吞吐量 96, 100
time 时间 53, 55
time complexity 时间复杂度 54, 59, 105, 112—113
tools 工具 18, 21, 24, 42, 122
top down approach 自上而下的方法 122
tractable problems 可处理的问题 59, 61

travelling salesman problem, the 旅行推销员问题 59—60
Turing, Alan 艾伦·图灵 xv, 25, 28, 113, 123
Turing machine 图灵机 13, 24, 25—29, 30, 60, 113
Turing test 图灵测试 114
Turing's thesis 图灵论题 28, 30
Turner, Mark 马克·特纳 126

U

Ullman, Jeffrey 杰弗里·乌尔曼 56
uncertainty 不确定性 5, 118
unconditional branch 无条件分支 74
uniprocessor architecture 单处理器架构 24
universal science 通用科学 129
universal Turing machine 通用图灵机 28
universe of discourse 话语空间 20
user-interface 用户界面 15, 17, 20, 24
utilitarian artefact 实用人工制品 53

V

variable 变量 71
verification and validation 验证和确认 79
very-short-term memory 非常短期存储器 87, 92
virtual machine 虚拟机 15, 21, 78, 85
virtual memory 虚拟存储 21
von Neumann, John 约翰·冯·诺伊

曼 81, 124

von Neumann style 冯·诺伊曼风格 25

W

Walter, Grey 格雷·沃尔特 2
water clock 水钟 2
Watson, James 詹姆斯·沃森 126
Watt, James 詹姆斯·瓦特 2
weak method 弱方法 115—116, 122
weight driven clock 重量驱动时钟 2

Whorf, Benjamin Lee 本杰明·李·沃尔夫 67
Wilkes, Maurice 莫里斯·威尔克斯 2, 4, 95, 102
Wing, Jeanette 周以真 120—123
word length 字长 88, 105
working memory 工作存储器 18, 87—88
world, artificial 人工世界 14
world, natural 自然世界 14
World Wide Web 万维网 15, 17, 20, 83
worst case performance 最坏情况 53, 55

Subrata Dasgupta
COMPUTER SCIENCE
A Very Short Introduction

*To
Anton Rippon*

Contents

Preface i

Acknowledgements v

List of illustrations vi

1 The 'stuff' of computing 1
2 Computational artefacts 13
3 Algorithmic thinking 33
4 The art, science, and engineering of programming 62
5 The discipline of computer architecture 81
6 Heuristic computing 104
7 Computational thinking 119

Epilogue: is computer science a universal science? 129

Further reading 133

Preface

The 1960s were tumultuous times, socially and culturally. But tucked away amidst the folds of the Cold War, civil rights activism, anti-war demonstrations, the feminist movement, student revolt, flower-power, sit-ins, and left-radical insurrections—almost unnoticed—a new science came into being on university campuses in the West and even, albeit more tentatively, in some regions of the non-Western world.

This science was centred on a new kind of machine: the electronic digital computer. The technology surrounding this machine was called by a variety of names, most commonly, 'automatic computation', 'automatic computing', or 'information processing'. In the English-speaking world this science was most widely called *computer science*, while in Europe it came to be labelled 'informatique', or 'informatik'.

The *technological idea* of automatic computation—designing and building real machines that would compute with minimal human intervention—can at least be traced back to the obsessive dreams of the English mathematician and intellectual gadfly Charles Babbage in the early 19th century, if not further back. The *mathematical concept* of computing was first studied in the late 1930s by the logicians Alan Turing in England and Alonzo Church in the United States. But the impetus for a proper *empirical science* of computing had to wait until the invention, design, and implementation of the electronic digital computer in the 1940s, just after the end of the Second World War. Even then, there was a gestation period. An autonomous science with a name and an identity of its own only emerged in the 1960s when universities began offering undergraduate and graduate degrees in computer science, and the first generation of formally trained *computer scientists* emerged from the campuses.

Since the advent of the electronic digital computer in 1946, the spectacular growth of the technologies associated with this machine (nowadays called generically 'information technology' or 'IT') and the related cultural and social transformation (expressed in such terms as 'information age', 'information revolution', 'information society') is visible for all to see and experience. Indeed, we are practically engulfed by this techno-social milieu. The *science*—the intellectual discipline—underlying the technology, however, is less visible and certainly less known or understood outside the professional computer science community. Yet computer science surely stands alongside the likes of molecular biology and cognitive science as being amongst the most consequential new sciences of the post-Second World War era. Moreover, there is a certain strangeness to computer science that compels attention and sets it apart from all other sciences.

My intent in this book is to offer the intellectually curious reader seriously interested in scientific ideas and principles the basis for an understanding of the fundamental nature of computer science; to enrich, if you will, the public understanding of this strange, historically unique, highly consequential, and still new, science. Put simply, this book strives to answer in direct, immediate, and concise fashion the question: *What is computer science?*

Before we proceed, some terminological clarity is in order. In this book I will use the word *computing* as a verb to denote a certain kind of activity; *computation* is used as a noun to signify the outcome of computing; *computational* is used as an adjective; *computer* is a noun which will refer to a device, artefact, or system that does computing; *artefact* refers to anything made by humans (or, sometimes, animals); and a *computational artefact* is any artefact that participates in computational work.

Finally, a caveat must be stated. This book begins by accepting the proposition that computer science is indeed a science; that is, it manifests the broad attributes associated with the concept of science, notably, that it entails the systematic blend of empirical, conceptual, mathematical and logical, quantitative and qualitative modes of inquiry into the nature of a certain kind of phenomena. Questioning this assumption is an exercise in the philosophy of science that is beyond the scope of this book. The abiding issue of interest here is the *nature* of computer science *qua* science and, especially, its distinct and distinguishing character.

Acknowledgements

I thank Latha Menon, my editor at OUP for her support and sage advice on this project from its very onset. Her comments on the penultimate version of the book were especially insightful.

Jenny Nugee always responded readily with editorial help and suggestions at various stages of this work. I thank her.

Four anonymous readers of two different drafts of the manuscript offered invaluable suggestions and comments which I took seriously. I am most grateful to them and wish I could acknowledge them by name.

My thanks to Elman Bashar for preparing the illustrations.

Portions of this material were used in an upper-level undergraduate course on 'Computational Thinking' which I have taught on several occasions to non-computer science majors. Their responses have been most helpful in shaping and sharpening the text.

Finally, as always, my thanks to members of my family. In their different ways each continues to provide the sustenance that makes living the life of the mind worthwhile.

List of illustrations

1 Abstraction and hierarchy inside a computer system **16**

2 General structure of the Turing machine **26**

3 Programming, related disciplines, and associated sciences **66**

4 Computer architectures and their external constraints **86**

5 Portrait of a computer's inner architecture **91**

6 An instruction pipeline **99**

7 Portrait of a multiprocessor **100**

8 General structure of a heuristic search system **115**

Chapter 1
The 'stuff' of computing

What is computer science? A, now classic, answer was offered in 1967 by three eminent early contributors to the discipline, Alan Perlis, Allen Newell, and Herbert Simon, who stated, quite simply, that computer science is the study of computers and their associated phenomena.

This is a quite straightforward response and I think most computer scientists would accept it as a rough and ready working definition. It centres on the computer itself, and certainly there would be no computer science without the computer. But both computer scientists and the curious layperson may wish to understand more precisely the two key terms in this definition: 'computers' and their 'associated phenomena'.

An automaton called 'computer'

The computer is an *automaton*. In the past this word, coined in the 17th century (plural, 'automata') meant any artefact which, largely driven by its own source of motive power, performed certain repetitive patterns of movement and actions without external influences. Sometimes, these actions imitated those of humans and animals. Ingenious mechanical automata have been devised since pre-Christian antiquity, largely for the amusement of the wealthy, but some were of a very practical nature as, for example, the water clock said to be invented in the 1st century CE by the engineer Hero of Alexandria. The mechanical weight-driven clock invented in 15th-century Italy is a highly successful and lasting descendant of this type of automaton. In the Industrial Revolution of the 18th century, the operation of a pump to remove water from mines motivated by the 'atmospheric' steam engine invented by Thomas Newcomen (in 1713), and later improved by James Watt (in 1765) and others, was another instance of a practical automaton.

Thus, mechanical automata that perform physical actions of one sort or another have a venerable pedigree. Automata that mimic cognitive actions are of far more recent vintage. A notable example is the 'tortoise' robot *Machina Speculatrix* invented by British neurophysiologist W. Grey Walter in the late 1940s to early 1950s. But the automatic electronic digital computer, developed in the second half of the 1940s, marked the birth process of an entirely new genus of automata; for the computer was an artefact designed to simulate and imitate certain kinds of human *thought* processes.

The idea of computing as a way of imitating human thinking—of the computer as a 'thinking machine'—is a profoundly interesting, disturbing, and controversial notion which I will address later in the book, for it is the root of a branch of computer science called *artificial intelligence* (AI). But many computer scientists prefer to be less anthropocentric about their discipline. Some even deny that computing has any similarity at all to autonomous human thinking. Writing in the 1840s, the remarkable English mathematician Ada Augustus, the Countess of Lovelace, an associate of Charles Babbage (see Preface) pointed out that the machine Babbage had conceived (called the Analytical Engine, the first incarnation of what a century later became the modern general purpose digital computer), had no 'pretensions' to initiating tasks on its own. It could only do what it was ordered to do by humans. This sentiment is often repeated by modern sceptics of AI, such as Sir Maurice Wilkes, one of the pioneers of the electronic computer. Writing at the end of the 20th century and echoing Lovelace, he insisted that computers only did what 'they had been written to do'.

So what *is* it that computers *do* which sets them apart from every other kind of artefact, including other sorts of automata? And what makes computer science so distinctive as a scientific discipline?

For the purpose of this chapter, I will treat the computer as a 'black box'. That is, we will more or less ignore the internal structure and workings of computers; those will come later. For the present we will think of the computer as a generic kind of automaton, and consider only *what* it does, not *how* it does what it does.

Computing as information processing

Every discipline that aspires to be 'scientific' is constrained by the fundamental *stuff* it is concerned with. The stuff of physics comprises matter, force, energy, and motion; that of chemistry is atoms and molecules; the stuff of genetics is the gene; and that of civil engineering comprises the forces that keep a physical structure in equilibrium.

A widely held view amongst computer scientists is that the fundamental stuff of computer science is *information*. Thus, the computer is the means by which information is automatically retrieved from the 'environment', stored, processed, or transformed, and released back into the environment. This is why an alternative term for computing is *information processing*; why in Europe computer science is called 'informatique' or 'informatik'; and why the 'United Nations' of computing is called the International Federation for Information Processing (IFIP).

The problem is that despite the founding of IFIP in 1960 (thus giving official international blessing to the concept of information processing), there remains, to this day, a great deal of misunderstanding about what information *is*. It is, as Maurice Wilkes once remarked, an *elusive* thing.

'Meaningless' information

One significant reason for this is the unfortunate fact that the word 'information' was appropriated by communication engineers to mean something very different from its everyday meaning. We usually think of information as telling us something *about* the world. In ordinary language, information is *meaningful*. The statement 'The average winter temperature in country X is 5 degrees Celsius' tells us something about the climate in country X; it gives us information about X. In contrast, in the branch of communication engineering called 'information theory', largely created by American electrical engineer Claude Shannon in 1948, information is simply a commodity transmitted across communication channels such as telegraph wires and telephone lines. In information theory, information is devoid of meaning. The unit of information in information theory is called the *bit* (short for 'binary digit') and a bit has only two values, usually denoted as '0' and '1'. However, in this age of personal computers and laptops, people are more familiar with the concept of the *byte*. One byte consists of eight bits. Since each bit can have one of two values, a byte of information can have 2^8 (= 256) possible values ranging from 00000000 to 11111111. What bits (or bytes) *mean* is of no concern in this sense of 'information'.

In computing, information processing in this meaningless sense is certainly relevant since (as we will see) a physical computer, made out of electronic circuits, magnetic and electromechanical devices, and the like (collectively dubbed 'hardware'), stores, processes, and communicates information as multiples of bits and bytes. In fact, one of the ways in which the capacity and performance of a computational artefact is specified is in terms of bits and bytes. For example, I may buy a laptop with 6 gigabytes of internal memory and 500 gigabytes of external memory ('hardrive'), (where 1 gigabyte = 10^9 bytes); or we may speak of a computer network transmitting information at the rate of 100 megabits/second (where 1 megabit = 10^6 bits).

'Meaningful' (or semantic) information

But the physical computer is (as we will see in Chapter 2) only one kind of computational artefact. Meaningless information is just one kind of information the computer scientist is interested in. The other, more significant (and arguably more interesting), kind is information that has meaning: *semantic* information. Such information connects to the 'real world'—and in this sense corresponds to the everyday use of the word. For example, when I access the Internet through my personal computer, information processing certainly occurs at the physical or 'meaningless' level: bits are transmitted from some remote computer ('server') through the network to my machine. But I am seeking information that is about something, say the biography of a certain person. The resulting text that I read on my screen means something to me. At this level, the computational artefact I am interacting with is a semantic information processing system.

Such information can, of course, be almost anything about the physical, social, or cultural environment, about the past, about thoughts and ideas of other people as expressed by them publicly, and even about one's own thoughts if they happen to be recorded or stored somewhere. What such meaningful information shares with meaningless information, as computer scientist Paul Rosenbloom has noted, is that it must be expressed in some physical medium such as electrical signals, magnetic states, or marks on paper; and that it resolves uncertainty.

Is information *knowledge*?

But consider an item of semantic information such as the biography of an individual. On reading it, I can surely claim to possess *knowledge* about that individual. And this points to the second source of confusion about the concept of information in ordinary language: the conflation of information with knowledge.

The poet T.S. Eliot had no doubt about their distinction. In his play *The Rock* (1934) he famously asked:

> Where is the wisdom we have lost in knowledge?
> Where is the knowledge we have lost in information?

Eliot was clearly implying a hierarchy: that wisdom is superior to knowledge, and knowledge to information.

Computer scientists generally avoid talking about wisdom as being beyond the scope of their purview. But they have also remained somewhat uneasy about distinguishing knowledge from information, at least in some contexts. For example, in AI, a subfield of computer science, a long-standing problem of interest has been knowledge *representation*—how to represent knowledge about the world in computer memory. Another kind of problem they study is how to *make inferences* from a body of knowledge. The kinds of things AI researchers recognize as knowledge include facts ('All men are mortal'), theories ('Evolution by natural selection'), laws ('Every action has an equal and opposite reaction'), beliefs ('There is a God'), rules ('Always come to a dead stop at a stop sign'), and procedures ('how to make seafood gumbo'), etc. But in what way such entities constitute knowledge and not information remains largely unsaid. AI researchers may well claim that what they do, in their branch of computer science, is knowledge processing rather than information processing; but they seem to fall shy of explaining why their concern is knowledge and not information.

In another specialty known as 'data mining' the concern is 'knowledge discovery' from large volumes of data. Some data mining researchers characterize knowledge as 'interesting' and 'useful' patterns or regularities hidden in large databases. They distinguish knowledge discovery from information retrieval (another kind of computing activity) in that the latter is concerned with retrieving 'useful' information from a database on the basis of some query, whereas the former identifies knowledge that is more than just 'useful' information, or more than patterns of regularity: such information must be 'interesting' in some significant sense to become knowledge. Like T.S. Eliot, data mining researchers rate knowledge as superior to information. At any rate, knowledge processing is what data mining is about rather than information retrieval.

Luciano Floridi, a philosopher of computing, offered the following view of the information/knowledge nexus. Information and knowledge bear a 'family resemblance'. They are both meaningful entities but they differ in that information elements are isolated like bricks whereas knowledge relates information elements to one another so that one can produce new inferences by way of the relationships.

To take an example: suppose, while driving, I hear on my car radio that physicists in Geneva have detected a fundamental particle called the Higgs boson. This new fact ('The Higgs boson exists') is certainly a piece of new information for me. I may even think that I have acquired some new knowledge. But this would be an illusion unless I can connect this information with other related items of information about fundamental particles and cosmology. Nor would I be able to judge the significance of this information. Physicists possess an integrated web of facts, theories, laws, etc., about subatomic particles, and about the structure of the universe that enable them to assimilate this new fact and grasp its significance or consequences. They possess the knowledge to do this, while I have merely acquired a new piece of information.

Is information *data*?

In mentioning 'data mining', I have introduced another term of great relevance: *data*. And here is yet another source of ambiguity in our making sense of the information concept, especially in the computer science community.

This ambiguity, indeed confusion, was remarked upon by the computer scientist Donald Knuth as far back as 1966, a time when computer science, emerging as a scientific discipline in its own right, was demanding the invention of new concepts and clarification of old ones. Knuth noted that in science there appeared to be some confusion concerning the terms 'information' and 'data'. When a scientist executes an experiment involving measurement, what is elicited might be any one of four entities: the 'true' values of that which is measured; the values that are actually obtained—approximations to the true values; a representation of the values; and the concepts the scientist teases out by analysing the measurements. The word 'data', Knuth asserted, most appropriately applies to the third of these entities. For Knuth, then, speaking as a computer scientist, data is the *representation* of information obtained by observation or measurement in some precise manner. So, in his view, information precedes data. In practice, the relationship between information and data is as murky as that between information and knowledge. Here, I can only cite a few of the diverse views of this relationship.

For Russell Ackoff, a prominent systems and management scientist, data constitute the outcome of observations; they are representations of objects and events. As for information, Ackoff imagined someone asking some questions *of* data which is then 'processed' (presumably by a human being or a machine) to afford answers, and this latter is information. So according to Ackoff, *contra* Knuth, data precedes information.

For Luciano Floridi, data also precedes information but in a different sense. Data exists, according to Floridi, only when there is an absence of uniformity between two states of a system. As he puts it, a datum (the infrequently used singular of 'data') exists whenever there are two variables, x and y such that $x \neq y$. So, for Floridi, data is a condition which itself has no meaning except that it signifies the presence of difference. When I am approaching a traffic light for instance, my observation of a red signal is a datum because it could have been otherwise: yellow or green.

Given this definition of data, Floridi then defines information as one or more data elements that are structured according to some rules, and are meaningful. To use the linguist's jargon, information is data when it possesses both syntax and semantics. Thus, my observation of the red traffic signal, a datum, becomes information because the meaning of the red light is that 'motorists must stop at the traffic light'. If I did not associate this action with the red light, the latter would remain only a datum.

As a final example, for AI researchers Jeffrey Shrager and Pat Langley, data do not result *from* observation; rather, observation *is* data; more precisely, what is observed is selectively recorded to qualify as data. Information does not figure in their scheme of things.

The programmer's point of view

These examples suffice to demonstrate the murkiness of the information/data connection from different perspectives. But let me return to Knuth. His definition of data reflects to a large extent, I think, the view of those computer scientists who specialize in another aspect of computer science, namely, computer programming—the techniques by which humans communicate a computational task to the computer (a topic I discuss later in this book). Even while paying lip service to the idea of computing as information processing, programmers and programming theorists do not generally reflect on 'information'; rather, they are more concerned with the Knuthian idea of data. More precisely, they concern themselves with data as the fundamental objects ('data objects') upon which computations are performed; and, thus, they are preoccupied with the classification of data objects ('data types'), the rules for representing complex data objects ('data structures'), and the rules for manipulating, processing, and transforming such data objects to produce new data objects. For such computer scientists it is data that matters, not information, not knowledge. To be more exact, programmers take for granted that there is information 'out there' in the 'real world'. But the interesting question for them is how to represent real world information in a form that is appropriate not only for automatic computing but also for human understanding. (Needless to say, other practitioners, such as historians, statisticians, and experimental scientists, do not usually regard data in this fashion.)

I will elaborate on this later in the book. But to give a very simple example of the programmer's view of data: in a university environment there will exist information in the registrar's office about its body of enrolled students: their names, dates of birth, home addresses, email addresses, names of parents or guardians, the subjects they are majoring in, the courses taken, the grades obtained, scholarships held, fees paid, and so on. The university administration needs a system that will organize this information in a systematic fashion (a 'database') such that, perhaps, information concerning any particular student can be accurately and speedily retrieved; new information about existing or new students can be inserted; the progress of individual students can be efficiently tracked; and statistics about the student population as a whole or some subset can be gathered. The programmer given the task of creating such a system is not concerned with the information per se, but rather, given the nature of the information, how to identify the basic data objects representing student information, construct data structures representing the data objects, and build a database so as to facilitate the computational tasks the university administration demands.

Symbol structures as the common denominator

I started this chapter with the proposition that the basic stuff of computing is information; that the computer is an automaton that processes information; and that consequently, computer science is the study of information processing.

But we have also seen that to some computer scientists (such as AI researchers) the fundamental stuff of computing is knowledge rather than information; and to others (such as programmers and programming theorists) it is data rather than information. We get a sense of the varied usages of these three entities from the following sample of terms found in the computing literature (some of which have already appeared in this chapter, others will be found in later ones):

> Data type, data object, data structure, database, data processing, data mining, big data…
>
> Information processing, information system, information science, information structure, information organization, information technology, information storage and retrieval, information theory…
>
> Knowledge base, knowledge system, knowledge representation, knowledge structure, theory of knowledge, declarative knowledge, procedural knowledge, knowledge discovery, knowledge engineering, knowledge level…

Can we then reduce these three entities, information, data, knowledge to a common denominator? Indeed we can. Computer scientist Paul Rosenbloom equated information with *symbols*, but we can go further. As far as computer science is concerned, all these three entities can be (and usually are) expressed by symbols—or, rather, by systems of symbols, *symbol structures*—that is, entities that 'stand for', represent, or denote other entities.

Symbols need a medium in which they are expressed, such as marks on paper. For example, the text 'The Higgs boson exists' is a symbol structure whose component symbols are alphabetic characters referring to sound units or phonemes, plus the 'blank' symbol; these when strung together represents something about the physical world. For the physics layperson, this is an item of information; for the particle physicist, this becomes a constituent of her knowledge system concerning fundamental particles. However, the physicist's knowledge which allows her to make sense of this information is itself a far more complex symbol structure stored in her brain and/or printed as text in books and articles. And Knuth's idea of data as the representation of information means that data are also symbol structures representing other symbol structures denoting information. Even the 'meaningless' information of information theory, the bits and bytes, are represented by physical symbols within a computer, such as voltage levels or magnetic states, or on paper by strings of 0s and 1s.

So in its *most* fundamental essence, the stuff of computing is symbol structures. *Computing is symbol processing.* Any automaton capable of processing symbol structures is a computer. The 'phenomena' associated with computers as Perlis, Newell, and Simon suggested are all ultimately reducible to symbol structures and their processing. Computer science is, ultimately, the science of automatic symbol processing, an insight which Allen Newell and Herbert Simon have emphasized. We may choose to call such symbol structures information, data, or knowledge depending on the particular 'culture' within computer science to which we belong.

It is this notion—that computing is ultimately symbol processing; that the computer is a symbol processing automaton; that computer science is the science of symbol processing—which sets computer science apart from other disciplines. As to its strangeness, this will be explained in a later chapter.

Chapter 2
Computational artefacts

We think of *the computer* as the centrepiece of computing; thus, of computer science. And rightly so. But there are caveats to be noted.

First, what exactly constitutes 'the computer' can be debated. Some tend to think of it as the physical object they work with on a daily basis (a laptop or their workplace desktop). Others think of the total system at their disposal, including such facilities as email service, word processing, accessing databases, etc., as 'the computer'. Still others relate it to an entirely mathematical model called the Turing machine (discussed later in this chapter).

Second, accepting that the computer is a symbol processing automaton, there are also other symbol processing artefacts associated with the computer, but which seem slightly at odds with our intuitive idea of 'the computer'. Thus, it behoves us to be more eclectic in our view of artefacts that participate in the computing process; hence the term *computational artefact*. In this chapter we consider the nature of computational artefacts.

In Chapter 1, the computer appeared as (more or less) a black box. All that was said was that it is a symbol processing automaton: it accepts symbol structures (denoting information, data, or knowledge as the case may be) as input and produces (of its own impetus) symbol structures as output.

When we prise open this black box we find that it is rather like a set of nested boxes: inside we find one or more smaller boxes; opening one of these inner boxes reveals still smaller boxes nested within. And so on. Of course, the degree of nesting of black boxes is finite; sooner or later we reach the most primitive boxes.

The natural and artificial worlds both manifest instances of this phenomenon—called *hierarchy*. Many physical, biological, social, and technological systems are hierarchical in structure. The difference between natural hierarchies (as in living systems) and artificial ones (as in cultural or technological systems) is that scientists have to *discover* the former and *invent* the latter.

The modern computer is a hierarchically organized system of computational artefacts. Inventing, understanding, and applying rules and principles of hierarchy is, thus, a subdiscipline of computer science.

There is a reason why hierarchies exist in both natural and artificial domains, and we owe this insight, most notably, to the polymath scientist Herbert Simon. Hierarchical organization, he stated, is a means of managing the *complexity* of an entity. In Simon's language, an entity is complex if it is composed of a number of components that interact in a non-trivial (that is, non-obvious) way. As we will see, the computer manifests this kind of complexity, hence it too is composed as a hierarchical system. The designers and implementers of computer systems are forced to structure them according to principles and rules of hierarchy. Computer scientists have the responsibility of inventing these rules and principles.

Compositional hierarchy

In general, a hierarchical system consists of components partitioned across two or more *levels*. The most common principles of hierarchy are concerned with the relationship of components both within and across levels.

Figure 1 depicts what I will call 'MY-COMPUTER'. (Physically, this may be a desktop, a laptop, a tablet, or even a smart phone. For convenience, I will assume it is one of the first two.) Suppose I use MY-COMPUTER only for three kinds of tasks: to write texts (as I am doing now), to send emails, and to search the (World Wide) Web via the Internet. Thus, I view it as consisting of three *computing tools* which I will call TEXT, MAIL, and WEB-SEARCH (level 1 of Figure 1), respectively. Each is a symbol processing computational artefact. Each is defined (for me as the tool user) in terms of certain *capabilities*. For example, TEXT offers a user-interface allowing me to input a stream of characters, and give commands to align margins, set spacing between lines, paginate, start a new paragraph, indent, insert special symbols, add footnotes and endnotes, italicize and boldface, and so on. It also allows me to input a stream of characters which, using the commands, is set into text which I can save for later use and retrieve.

From my point of view, TEXT *is* MY-COMPUTER when I am writing an article or a book (as at this moment), just as MAIL or WEB-SEARCH *is* MY-COMPUTER when I am emailing or searching the Web, respectively. More precisely, I am afforded three different, alternative *illusions* of what MY-COMPUTER is. Computer scientists refer to such illusionary artefacts as *virtual machines*, and the creation, analysis, and understanding of such virtual machines is one of the major concerns in computer science. They constitute one of the phenomena surrounding computers that Perlis, Newell, and Simon alluded to.

The term *architecture* is used generically by computer scientists to mean the logical or functional structure of computational artefacts. (The term *computer architecture* has a more specialized meaning which will be discussed later.) From my (or any other user's) point of view, the computing tool TEXT has a certain architecture which is visible to me: it has an interpreter that interprets and executes commands; a temporary or working

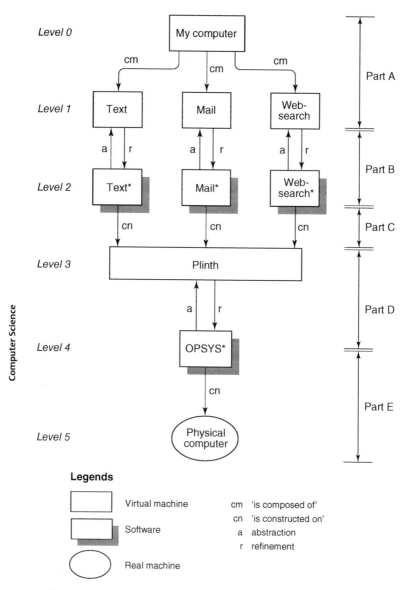

1. Abstraction and hierarchy inside a computer system.

memory whose content is the text I am composing; a permanent or long-term memory which holds all the different texts as files I have chosen to save; input channels that transmits my character streams and commands to the machine, and output channels that allow the display of texts on a screen or as printed matter ('hardcopy'). These components are 'functional': I may not know (or particularly care) about the actual media in which these components exist. And because they characterize all I (as the user) need to know about TEXT to be able to use it, we will call it TEXT's architecture.

Likewise, when emailing, the tool MAIL *is* MY-COMPUTER: a virtual computational artefact. This too is a symbol processor. It manifests a user-interface that enables me to specify one or more recipients of a message; link one or more other symbol structures (texts, pictures) as attachments that accompany the message; compose the message; and send it to the recipient(s). Its architecture resembles that of TEXT in that it manifests the same kinds of components. It can interpret commands, has an input channel enabling character streams to be assembled in a working memory, a long-term memory to hold, as long as I want, my messages, and output channels for displaying the contents of the email on the screen and printing it out. In addition, MAIL has access to other kinds of long-term memory which hold the symbol structures (texts and images) that can be attached to the message; however, one of these long-term memories is *private* to MY-COMPUTER and so, can only be accessed by me, while the other is *public*—that is, shared with users of other computers.

Finally, there is WEB-SEARCH. Its architecture is similar: an interpreter of the commands; a shared/public memory (the Web) whose contents ('web pages') are accessible; a private working memory that (temporarily) holds the contents accessed from shared memory; a private long-term memory which can save these contents; and input and output channels.

The hierarchy shown in part A of Figure 1 is two-levelled. At the upper level (0) is a single computational artefact, MY-COMPUTER; but the lower level (1) shows that MY-COMPUTER is composed of three independent tools. This lower level constitutes my tool box as it were. This type of hierarchy, when an entity A is composed of entities $α, β, γ, \ldots$, is ubiquitous in complex systems of any kind, natural or artificial. It is certainly a characteristic of computational artefacts. There is no commonly accepted term for it, so let us call it *compositional hierarchy*.

Abstraction/refinement

As we have noted, the three computing tools at level 1 of Figure 1 are similar in their respective architectures. Each comprises of shared and private long-term memories, private working memory, input and output channel(s), and interpreter(s) of commands.

But these three computing tools must have been *implemented* for them to be actual working artefacts: For instance, someone must have designed and implemented a computational artefact which when activated *performs* as TEXT, hiding the details of the mechanisms by which TEXT was realized. Let us denote this implemented artefact TEXT* (level 2 of Figure 1); this is a computer program, a piece of software. The relationship between TEXT and TEXT* is one of abstraction/refinement (part B of Figure 1):

> An *abstraction* of an entity E is itself another entity e that reveals only those characteristics of E considered relevant in some context while suppressing other characteristics deemed irrelevant (in that context). Conversely, a *refinement* of an entity e is itself another entity E such that E reveals characteristics that were absent or suppressed in e.

TEXT is an abstraction of TEXT*; conversely, TEXT* is a refinement of TEXT. Notice that abstraction/refinement is also a principle of hierarchy in which abstraction is at the upper level and refinement at the lower level. Notice also that abstractions and refinements are context-dependent. The same entity E may be abstracted in different ways to yield two or more higher level entities $e1, e2, \ldots, eN$. Conversely, the same entity e may be refined in two more different ways to yield different lower level entities $E1, E2, \ldots, En$.

The principle of abstraction/refinement as a way of managing the complexity of computational artefacts has a rich history from the earliest years of computing. Perhaps the person who made the emerging computer science community in the 1960s most conscious about the importance of this principle of hierarchy was Edsger Dijkstra. Later we will see its particular importance in the process of *building* computational artefacts, but for the present it is enough for the reader to appreciate how a complex computational artefact can be *understood* in terms of the abstraction/refinement principle, just as a complex computational artefact can be understood in terms of the compositional hierarchy.

Hierarchy by construction

We are no longer 'seeing' MY-COMPUTER from my perspective as a user. We are now in the realm of those who have actually created MY-COMPUTER: the tool builders or artificers. And they are not, incidentally, a homogenous lot.

In particular, TEXT*, MAIL*, and WEB-SEARCH* are computer programs—*software*—which programmers (software developers, as they are now preferably called), have constructed upon an infrastructure which I call here PLINTH (part C and level 3 of Figure 1). This particular infrastructure also consists of a collection of computing tools the builders of TEXT* et al. could use. Here, then, is a third type of hierarchy: *hierarchy by construction*.

Ever since the earliest days of the digital computer, designers and researchers have sought to protect the user, as far as possible, from the gritty, and sometimes nasty, realities of the physical computer, to make the user's life easier. The ambition has always been to create a smooth, pleasant user interface that is close to the user's particular universe of discourse, and remains within the user's comfort zone. A civil or a mechanical engineer wants to perform computations that command the computer to solve the equations of engineering mechanics; a novelist wants his computer to function as a writing instrument; the accountant desires to 'offload' some of her more tedious calculations to the computer; and so on. In each case, the relevant user desires an *illusion* that their computer is tailor-made for his or her need. Much debate has ensued over the years as to whether such user tools and infrastructures should be incorporated into the physical machine ('hardwired') or provided in more flexible fashion by way of software. In general, the partitioning of infrastructures and tools across this divide has been rationalized by the particular needs of the communities for whom computers are developed.

As we have noted, MY-COMPUTER offers such illusions to the user whose sole concerns are typing text, sending and receiving emails, and searching the Web. MY-COMPUTER offers an infrastructure for the user to write texts, compose and send emails, and search for information on the Web, just as PLINTH (at a lower level) offers such an infra-structure for the construction of programs that function as the user's toolkit.

But even the software developers, who created these abstractions by implementing the programs TEXT* et al., must have their own illusions: they too are users of the computer though their engagement with the computer is far more intense than mine when I am using TEXT or MAIL. We may call them 'application programmers' or 'application software developers', and they too must be shielded from some of the realities of the physical computer. They too need an infrastructure *with* which they can work, *upon* which they can create their own virtual machines.

In Figure 1, the entity named PLINTH is such a foundation. It is, in fact, an abstraction of a collection of programs (a 'software system'), shown here as OPSYS* (level 4) which belongs to a class of computational artefacts called *operating systems*.

An operating system is the great facilitator; it is the great protector; it is the great illusionist. In its early days of development, in the 1960s, it was called 'supervisor' or 'executive' and these terms capture well what its responsibilities are. Its function is to manage the resources of the physical computer and provide a uniform set of services to all users of the computer whether layperson or software developer. These services include 'loaders' which will accept programs to be executed and allocating them to appropriate locations in memory; memory management (ensuring that one user program does not encroach upon, or interfere with, the memory used by another program); providing *virtual* memory (giving users the illusion of unlimited memory); controlling physical devices (such as disks, printers, monitors) that perform input and output functions; organizing the storage of information (or data or knowledge) in long-term memories so as to make it easily and speedily accessible; executing procedures according to standardized rules (called 'protocols') that enable a program on one computer to request service from a program in another computer communicated through a network; protecting a user's program from being corrupted by another user's program either accidentally or by the latter user's malice. The infrastructure called PLINTH in Figure 1 provides such services—a set of computing tools; it is an abstraction of the operating system OPSYS*.

Yet, an operating system is not exactly a firewall forbidding all interaction between a program constructed atop it (such as

MAIL*) and the physical computer beneath it. After all, a program will execute by issuing instructions or commands to the physical computer, and most of these instructions will be directly interpreted by the physical computer (in which situation, these instructions are called 'machine instructions'). What the operating system will do is 'let through' machine instructions to the physical computers in a controlled fashion, and interpret other instructions itself (such as those for input and output tasks).

Which brings us to (almost) the bottom of the hierarchy depicted in Figure 1. OPSYS*, the operating system software, is shown here as *constructed*—on top of the physical computer (level 5). For the present we will assume that the physical computer (commonly and crudely called *hardware*) is (finally) the 'real thing'; that there is nothing virtual about it. We will see that this too is an illusion, that the physical computer has its own internal hierarchy and it too has its own levels of abstraction, composition, and construction. But at least we can complete the present discussion on this note: that the physical computer provides an infrastructure and a toolbox comprising a repertoire of instructions (machine instructions), a repertoire of data types (see Chapter 1), modes of organizing and accessing instructions and data in memory, and certain other basic facilities which enable the implementation of programs (especially the operating system) that can be executed by the physical computer.

Three classes of computational artefacts

In a recent book narrating the history of the birth of computer science I commented that a peculiarity of computer science lies in its three classes of computational artefacts.

One class is *material*. These artefacts, like all material objects encountered through history, obey the physical laws of nature (such as Ohm's law, the laws of thermodynamics, Newton's laws of motion, etc.). They consume power, generate heat, entail (in some cases) physical motion, decay physically and chemically over time, occupy physical space, and consume physical time when operational. In our example of Figure 1, the physical computer at level 5 is an instance. Obviously, all kinds of computer hardware are material computational artefacts.

Some computational artefacts, however, are entirely *abstract*. They not only process symbol structures, they *themselves* are symbol structures and are intrinsically devoid of any physicality (though they may be made visible via physical media such as marks on paper or on the computer screen). So physico-chemical laws do not apply to them. They neither occupy physical space nor do they consume physical time. They 'neither toil nor spin' in physical space-time; rather, they exist in their own space-time frame. There are no instances of the abstract artefact in Figure 1. In the next section, I cite examples, and will discuss some of them in chapters to follow. But if you recall the mention of procedures that I as a user of TEXT or MAIL can devise to deploy these tools, such procedures exemplify abstract artefacts.

The third class of computational artefacts are the ones that most lend *strangeness* to computer science. These are abstract *and* material. To be more precise, they are themselves symbol structures, and in this sense they are abstract; yet their operations cause changes in the material world: signals transmitted across communication paths, electromagnetic waves to radiate in space, physical states of devices to change, and so on; moreover, their actions depend on an underlying material agent to execute the actions. Because of this nature, I have called this class *liminal* (meaning a state of ambiguity, of between and betwixt). Computer programs or software is one vast class of liminal computational artefacts, for example, the programs TEXT*, MAIL*, WEB-SEARCH*, and the operating system OPSYS* of Figure 1.

Later, we will encounter another important kind of liminal artefact. For the present, what makes computer science both distinctive and strange is not only the presence of liminal artefacts but also that what we call 'the computer' is a *symbiosis* of the material, the abstract, and the liminal.

Over the approximately six decades during which computer science as an autonomous, scientific discipline evolved, many distinct subclasses of these three classes of computational artefacts have emerged. Four instances—user tool and infrastructure, software, and physical computer—are shown in Figure 1. Of course, some subclasses are more central to computing than others because they are more *universal* in their scope and use than others. Moreover, the classes and subclasses form a compositional hierarchy of their own.

Here is a list of some of these classes and subclasses presently recognized in computer science. The numbering convention demonstrates the hierarchical relationship between them. While the reader may not be familiar with many of these elements, I will explain the most prominent of them in the course of this book.

[1] Abstract artefacts
    [1.1] Algorithms
    [1.2] Abstract automata
        [1.2.1] Turing machines
        [1.2.2] Sequential machines
    [1.3] Metalanguages
    [1.4] Methodologies
    [1.5] Languages
        [1.5.1] Programming languages
        [1.5.2] Hardware description languages
        [1.5.3] Microprogramming languages
[2] Liminal artefacts
    [2.1] User tools and interfaces
    [2.2] Computer architectures
        [2.2.1] Uniprocessor architectures
        [2.2.2] Multiprocessor architectures

[2.2.3] Distributed computer architectures
[2.3] Software (programs)
[2.3.1] Von Neumann style
[2.3.2] Functional style
[3] Material artefacts
[3.1] Physical computers/hardware
[3.2] Logic circuits
[3.3] Communication networks

The 'great unifier'

There is one computational artefact that must be singled out. This is the *Turing machine*, an abstract machine named after its originator—logician, mathematician, and computer theorist Alan Turing. Let me first describe this artefact and then explain why it deserves special attention.

The Turing machine consists of a tape that is unbounded in length and divided into squares. Each square can hold one of a vocabulary of symbols. At any point in time a read/write head is positioned on one square of the tape which becomes the 'current' square. The symbol in the current square (including the 'empty' symbol or 'blank') is the 'current symbol'. The machine can be in one of a finite number of *states*. The state of the machine at any given time is its 'current state'. Depending on the current symbol and the current state, the read/write head can write (an output) symbol on the current square (overwriting the current symbol), move one square left or right, or effect a change of state, called the 'next state'. The cycle of operation repeats with the next state as the current state, the new current square holding the new current symbol. The relationships between the (possible) current states (CS), (possible) current (input) symbols (I), the (possible) output symbols (O), movements of the read/write head (RW), and the (possible) next states (NS) are specified by a 'state table'. The behaviour of the machine is controlled by the state table and the invisible mechanism that effects the reads and writes, moves the read/write head, and effects changes of state.

Figure 2 depicts a very simple Turing machine which reads an input string of 0s and 1s written on the tape, replaces the input string with 0s except that when the entire string has been scanned, it writes a 1 if the number of 1s in the input string is odd, and 0 otherwise. The machine then comes to a *halt*. A special symbol, say #, on the tape indicates the end of the input string. This machine would be called a 'parity detector': it replaces the entire input string with 0s and replaces # with a 1 or a 0 depending on whether the parity of (the number of 1s in) the input string is odd or even.

This machine needs three states: *So* signifies that an odd number of 1s have been detected in the input string at any point in the machine's operation. *Se* represents the detection of an even number of 1s up to any point in the machine's operation. The third state *H* is the halting state: it causes the machine to halt. When the machine begins operation, its read/write head is pointing to the square holding the first digit in the input string.

The potential behaviour of the Turing machine is specified by the state table (see Table 1).

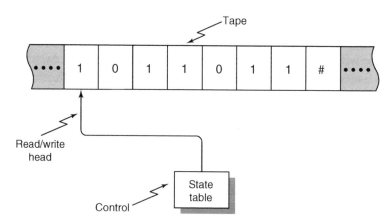

2. **General structure of the Turing machine.**

Table 1. The state table

Current state	Input symbol	Next state	Output symbol	Move read/write head
Se	0	Se	0	R
Se	1	So	0	R
Se	#	H	0	—
So	0	So	0	R
So	1	Se	0	R
So	#	H	1	—

Each row in this table specifies a distinct operation on the part of the machine and must be interpreted independently. For example, the first row says that: *if* the current state is *Se and* the current input symbol is 0 *then* the next state will (also) be *Se and* output symbol 0 is written on the tape *and* the read/write head is moved one position right. The last row tells us that that *if* the current state is *So and* the input symbol is # *then* replace the # with a 1 *and make* the next state the halt state *H*. There is no further motion of the read/write head.

Suppose the input string is as shown in Figure 2, and the machine is set to the state *Se*. The reader can easily verify that the sequence of states and the contents of the tape in successive cycles of the machine's operation will be as follows. The position of the read/write head in each cycle is indicated by the asterisk to the right of the 'current' input symbol:

Se: 1*011011# → *So*: 00*11011# → *So*: 001*1011# → *Se*: 0001*011# →
So: 00000*11# → *So*: 000001*1# → *Se*: 0000001*# → *So*: 0000000#* →
H: 00000001*

There will be, then, a distinct Turing machine (Turing himself called this, simply, a 'computing machine') for each distinct symbol processing task. Each such (special purpose) Turing machine will specify the alphabet of symbols that the machine will recognize, the set of possible states, the initial square on which the read/write head is positioned, the state table, and the initial current state. At the end of the machine's operation (when it reaches the 'halt' state, if there is one) the output written onto the tape gives the result of the symbol processing task.

Thus, for example, a Turing machine can be built to add two numbers n, m, represented by n 1s followed by a blank followed by m 1s, leaving the result $n + m$ (as a string of $n + m$ 1s) on the tape. Another Turing machine with a single string composed of the symbols a, b, and c as input will replace this input string with a 'mirror image' (called a 'palindrome') of the input string. For example if the input string is '*aaabbbccc*' then the output will be '*cccbbbaaa*'. A Turing machine is, thus, a symbol processing machine. It is, of course, an abstract artefact in the 'purest' sense since the machine itself is a symbol structure. No one would dream of making a physical version of a Turing machine as a *practical* artefact.

But Turing went further. He also showed that one can build a single computing machine **U** that can *simulate* every other Turing machine. If **U** is provided with a tape containing the description of the state table for a specific Turing machine, **U** will interpret that description and perform the same task as that particular machine would do. Such a machine **U** is called a *universal Turing machine*.

The significance of Turing's invention lies in a claim he made that *any procedure that we 'intuitively' or 'naturally' think of as a computing procedure can be realized by a Turing machine*. It follows that a universal Turing machine can perform anything we think of as computing. This claim is called the *Turing thesis* (or sometimes as the Church–Turing thesis, since another logician, Alonzo Church, arrived at the same conclusion using an entirely different line of thinking).

We may think of the Turing machine as the 'great unifier'. It is what binds all computational artefacts; that is, all computational artefacts and their behaviours can be *reduced* to the workings of a Turing machine.

Having said this, and also recognizing that an entire branch of computer science called *automata theory* exists which studies the structure and behaviour, the power and limitations of the Turing machine in all its conceivable manifestations (e.g. in confining the tape to a finite length, or in introducing multiple tapes with multiple read/write heads), we must also recognize the paradoxical situation that the Turing machine has had almost no impact on the invention, design, implementation, and behaviour of any practical (or practicable) computational artefact whatsoever, or on the thinking and practice of computer scientists who deal with such artefacts!

Interactive computing

Moreover, since Turing's time there have emerged computational artefacts that work *interactively* with each other or with other natural or artificial systems. 'Interaction' refers here to the mutual or reciprocal influence amongst artificial (including social) and/or natural *agents* that together form a system of some sort.

Consider, for example, my paying a utilities bill: this entails an interaction between me and my laptop, and my bank's computer system and that of the utility company. In this situation four agents (three computational artefacts and myself) are effecting information transfers and computations interactively, by exchanging messages, commands, and data.

Or consider the abstract computational artefact TEXT in Figure 1. This constitutes a *human–computer interface* whereby the human user of TEXT and the software system TEXT* interact with each other. Commands afforded by TEXT and issued by the user causes

TEXT* to respond (initiating a new line of text, creating a space between words, adding characters to form words in the text, indenting for a new paragraph, italicizing a word, etc.) and this latter response prompts, in turn, the human user to respond.

Such interactive systems do not conform to the 'standard' idea of the Turing machine which is essentially a stand-alone artefact with inputs already inscribed on its tape before the activation of the machine and whose output is only visible when the Turing machine halts. Interactive computational artefacts (such as my bank's or my utility company's system) may never halt.

It is because of such considerations that some computer scientists insist that the study of Turing machines—automata theory—properly belongs to the realm of mathematics and mathematical logic than to computer science proper, while others question the validity of seeing the Turing thesis as encompassing the whole of computing.

Computer science as a science of the artificial

To summarize the discussion so far, computational artefacts are *made* things; they process symbol structures signifying information, data, or knowledge (depending on one's point of view and context). Computer science is the science of computational artefacts.

Clearly, computational artefacts are not part of the natural world in the sense that rocks, minerals and fossils, plants and animals, stars, galaxies and black holes, elementary particles, atoms and molecules are. Human beings bring these artefacts into existence. Thus, computer science is not a *natural* science. So what kind of science is it?

One view is that since computational artefacts are utilitarian, thus technological, computer science is not 'really' a science at all. Rather, it is a branch of engineering. However, the traditional engineering sciences such as strength of materials, theory of structures, thermodynamics, physical metallurgy, circuit theory, as well as such new engineering sciences as bioengineering and genetic engineering are directly constrained by the laws of nature. Liminal and abstract computational artefacts seem a far cry from the uncompromisingly material artefacts—structures, machine tools, engines, integrated circuits, metals, alloys, and composite materials, etc.—studied by engineering scientists. This is one of the reasons why material computational artefacts (computer hardware) often belong to the domain of engineering schools while liminal and abstract ones are in the domain of schools of science.

However, all artefacts—engineering and computational—have something in common: they are the products of human thought, human goals, human needs, human desires. *Artefacts are purposive: they reflect the goals of their creators.*

Herbert Simon called all the sciences concerned with artefacts (abstract, liminal, or material) the *sciences of the artificial*. They stand apart from the natural sciences because they must take into account goals and purposes. A natural object has no purpose: rocks and minerals, stars and galaxies, atoms and molecules, plants and organisms have not come into the world with a purpose. They just *are*. The astronomer does not ask: 'What is a galaxy for?' The geologist does not ask: 'What is the purpose of an igneous intrusion?' The task of the natural scientist is to discover the laws governing the structures and behaviours of natural phenomena, inquire into how they came into being, but not ask why—for what purpose—they came into existence.

In contrast, artefacts have entered the world reflecting human needs and goals. It is not enough to ask what are the laws and principles governing the structure and behaviour of a computational artefact (or, for that matter, of pyramids, suspension bridges, particle accelerators, and kitchen knives) if we then ignore the reason for their existence.

The sciences of the artificial entail the study of the *relationship between means and ends*: the goals or needs for which an artefact is intended, and the artefact made to satisfy the needs. The 'science' in computer science is, thus, a science of means and ends. It asks: given a human need, goal, or purpose, how can a computational artefact *demonstrably* achieve such a purpose? That is, how can one demonstrate, by reason or observation or experiment that the computational artefact satisfies that purpose?

Chapter 3
Algorithmic thinking

Like the character in Molière's play who did not know he had been speaking prose all his life, most people may not realize that, when as children they first multiplied two multi-digit numbers or did long division, they were executing an *algorithm*. Indeed, it is probably the case that before the 1960s few people outside the computing and mathematical communities knew the word 'algorithm'. Since then, however, like 'paradigm' (a term originally made fashionable in the rarified reaches of philosophy of science) 'algorithm' has found its way into common language to mean formulas, rules, recipes, or systematic procedures to solve problems. This is largely due to the intimate association, in the past five decades or so, of computing with algorithms.

Yet, the *concept* of an algorithm (if not the word) reaches back to antiquity. Euclid's great work, *Elements* (*c*.300 BCE) where the principles of plane geometry were laid out, described an algorithm to find the greatest common divisor (GCD) of two positive integers. The word 'algorithm' itself originated in the name of 9th-century Arabic mathematician and astronomer, Mohammed ibn-Musa al-Khwarizmi, who lived and worked in one of the world's premier scientific centres of his age, the House of Wisdom in Baghdad. In one of his many treatises on mathematics and astronomy, al-Khwarizmi wrote on the 'Hindu art of reckoning'. Mistaking this work as al-Khwarizmi's own, later readers of Latin translations called his work 'algorismi' which eventually became 'algorism' to mean a step-by-step procedure. This metamorphosed into 'algorithm'. The earliest reference the *Oxford English Dictionary* could find for this word is in an article published in an English scientific periodical in 1695.

Donald Knuth (who perhaps more than any other person made algorithms part of the computer scientist's consciousness) once described computer science as the study of algorithms. Not all computer scientists would agree with this 'totalizing' sentiment, but none could conceive of a computer science without algorithms at its epicentre. Much like the Darwinian theory of evolution in biology, all roads in computing seem to lead to algorithms. If to think biologically is to think evolutionarily, to think computationally is to form the habit of algorithmic thinking.

The litmus test

As an entry into this realm, consider the *litmus test*, which is one of the first experiments a student performs in high school chemistry.

There is a liquid of some unknown sort in a test tube or beaker. The experimenter dips a strip of blue litmus paper into it. It turns red, therefore the liquid is acidic; it remains blue, it is not acidic. In the latter case the experimenter dips a red litmus strip into the liquid. It turns blue, therefore the liquid is alkaline (basic), otherwise it is neutral.

This is a *decision procedure* chemistry students learn very early in their chemical education, which we can describe in the following fashion:

    if a blue litmus strip turns red when dipped into a liquid
        then conclude the liquid is acidic
        else
            if a red litmus strip turns blue when dipped in the liquid

> **then** conclude the liquid is alkaline
> **else** conclude the liquid is neutral

The notation used here will appear throughout this chapter and in some of the chapters that follow, and needs to be explained. In general, the notation **if** C **then** $S1$ **else** $S2$, is used in algorithmic thinking to specify decision making of a certain sort. If the condition C is true then the *flow of control in the algorithm* goes to the segment $S1$, and $S1$ will then 'execute'. If C is false then control goes to $S2$, and $S2$ will then execute. In either case, after the execution of the **if then else** statement control goes to the statement that follows it in the algorithm.

Notice that the experimenter does not need to know anything about *why* the litmus test works the way it does. She does not need to know what 'litmus' actually is—its chemical composition—nor what chemical process occurs causing the change of colour. To carry out the procedure it is entirely sufficient that (a) the experimenter recognizes litmus paper when she sees it; and (b) she can associate the changes of colour with acids and bases.

The term 'litmus test' has become a metaphor for a definitive condition or test. And for good reasons: it is guaranteed to work. There *will* be an unequivocal outcome; there is no room for uncertainty. Moreover, the litmus test cannot go on indefinitely; the experimenter is assured that within a *finite amount of time* the test will give a decision.

These conjoined properties of the litmus test—a mechanical procedure which is guaranteed to produce a correct result in a finite amount of time—are essential elements characterizing an algorithm.

When is a procedure an algorithm?

For computer scientists, an algorithm is not just a mechanical procedure or a recipe. In order for a procedure to qualify as an algorithm as computer scientists understand this concept, it must possess the following attributes (as first enunciated by Donald Knuth):

Finiteness. An algorithm always terminates (that is, comes to a halt) after a finite number of steps.

Definiteness. Every step of an algorithm must be precisely and unambiguously specified.

Effectiveness. Each operation performed as part of an algorithm must be primitive enough for a human being to perform it exactly (using, say, pencil and paper).

Input and output. An algorithm must have one or more inputs and one or more outputs.

Let us consider Euclid's venerable algorithm mentioned earlier to find the GCD of two positive integers m and n (that is, the largest positive integer that divides exactly into m and n). The algorithm is described here in a language that combines ordinary English, elementary mathematical notation, and some symbols used to signify decisions (as in the litmus test example). In the algorithm, m and n serve as 'input variables' and n also serves as an 'output variable'. In addition, a third 'temporary variable', denoted as r is required. A 'comment', which is not part of the algorithm itself is enclosed in '{ }'. The '←' symbol in the algorithm is of special interest: it signifies the 'assignment operation': '$b \leftarrow a$' means to *copy* or *assign* the value of the variable a into b.

Euclid's GCD algorithm. Given two positive integers m and n find their GCD.
Input m, n {$m, n \geq 1$};
Temp var r;
Step 1: divide m by n; $r \leftarrow$ remainder; {$0 \leq r \leq$};
Step 2: if $r = 0$ **then Output** n;
Halt

       else
Step 3: $m \leftarrow n$;
   $n \leftarrow r$;
   goto step 1.

Suppose initially $m = 16$, $n = 12$. If a person 'executes' this algorithm using pencil and paper, then the values of the three variables m, n, r after each step's execution will be as follows.

	m	n	r
Step 1:	16	12	4;
Step 3:	12	4	4;
Step 1:	12	4	0;
Step 2:	Output $n = 4$.		

As another example, suppose initially $m = 17$, $n = 14$. The values of the three variables after each step's execution will be the following:

	m	n	r
Step 1:	17	14	3;
Step 3:	14	3	3;
Step 1:	14	3	2;
Step 3:	3	2	2;
Step 1:	3	2	1;
Step 3:	2	1	1;
Step 1:	2	1	0;
Step 2:	Output $n = 1$.		

In the first example, GCD (16, 12) = 4, which is the algorithm's output when it halts; in the second example GCD (17, 14) = 1, which the algorithm outputs after it terminates.

Clearly, the algorithm has inputs. Much less obvious is whether the algorithm satisfies the finiteness criterion. There is a repetition or *iteration* indicated by the **goto** command which causes control to return to step 1. As the two examples indicate, the algorithm iterates between steps 1 and 3 until the condition $r = 0$ is satisfied whereupon the value of n is output as the result and the algorithm comes to a halt. The two examples indicate clearly that for these *particular* pairs of input values for m and n, the algorithm always ultimately satisfies the *termination criterion* ($r = 0$) and will halt. However, how do we know whether for other pairs of values it will not iterate forever alternating steps 1 and 3 and never produces an output? (Thus, in that situation, the algorithm will violate both the finiteness and output criteria.) How do we know that the algorithm will always terminate for all possible positive values of m and n?

The answer is that it must be *demonstrated* that in general, the algorithm is finite. This demonstration lies in that after every test of $r = 0$ in step 2, the value of r is less than the positive integer n, and the values of n and r decreases with every execution of step 1. A decreasing sequence of positive integers must eventually reach 0 and so eventually $r = 0$, and so by virtue of step 2 the procedure will eventually terminate.

What about the definiteness criterion? What this says is that every step of an algorithm must be precisely defined. The actions to be carried out must be unambiguously specified. Thus *language* enters the picture. The description of Euclid's algorithm uses a mix of English and vaguely mathematical notation. The person who mentally executes this algorithm (with the aid of pencil and paper) is supposed to understand exactly what it means to divide, what a remainder is, what positive integers are. He must understand the meaning of the more formal notation, such as the symbols '**if...then...else**', '**goto**'.

As for effectiveness, all the operations to be performed must be primitive enough that they can be done in a finite length of time. In this particular case the operations specified are basic enough that one can carry them out on paper as was done earlier.

'Go forth and multiply'

The concept of *abstraction* applies to the specification of algorithms. In other words, a particular problem may be solved by algorithms specified at two or more different levels of abstraction.

Before the advent of pocket calculators children were taught to multiply using pencil and paper. The following is what I was taught as a child. For simplicity, assume a three-digit number (the 'multiplicand') is being multiplied by a two-digit number (the 'multiplier').

> **Step 1**: Place the numbers so that the multiplicand is the top row and the multiplier is the row below and position them so that the units digit of the multiplier is aligned exactly below the units digit of the multiplicand.
>
> **Step 2**: Draw a horizontal line below the multiplier.
>
> **Step 3**: Multiply the multiplicand by the units digit of the multiplier and write the result ('partial product') below the horizontal line, positioning it so that the units digits are all aligned.
>
> **Step 4**: Place a '0' below the units digit of the partial product obtained in step 3.
>
> **Step 5**: Multiply the multiplicand with the tens digit of the multiplier and position the result (partial product) on the second row below the horizontal line to the left of the '0'.
>
> **Step 6**: Draw another horizontal line below the second partial product.
>
> **Step 7**: Add the two partial products and write it below the second horizontal line.
>
> **Step 8**: Stop. The number below the second line is the desired result.

Notice that to perform this procedure successfully the child has to have *some* prior knowledge: (a) She must know how to multiply a multi-digit number by a one-digit number. This entails either memorizing the multiplication table for two one-digit multiplications or having access to the table. (b) She must know how to add two or more one-digit numbers; and she must know how to handle carries. (c) She must know how to add two multi-digit numbers.

However, the child does *not* need to know or understand *why* the two numbers are aligned according to step 1; or *why* the second partial product is shifted one position to the left as per step 5; or *why* a '0' is inserted in step 4; or *why* when she added the two partial products in step 7 the correct result obtains.

Notice, though, the preciseness of the steps. As long as someone follows the steps exactly as stated this procedure is guaranteed to work provided conditions (a) and (b) mentioned earlier are met by the person executing the procedure. It is guaranteed to produce a correct result in a finite amount of time: the fundamental characteristics of an algorithm.

Consider how most of us nowadays would do this multiplication. We will summon our pocket calculator (or smart phone), and we will proceed to use the calculator as follows:

Step 1': Enter the multiplicand.
Step 2': Press 'x'.
Step 3': Enter the multiplier.
Step 4': Press '='.
Step 5': Stop. The result is what is displayed.

This is also a multiplication algorithm. The two algorithms achieve the same result but they are at two different levels of abstraction. Exactly what happens in executing steps 1'–4' in the second algorithm the user does not know. It is quite possible that the calculator is *implementing* the same algorithm as the paper and pencil version. It is equally possible that a different implementation is used. This information is hidden from the user.

The levels of abstraction in this example also imply levels of *ignorance*. The child using the paper-and-pencil algorithm knows *more* about multiplication than the person using a pocket calculator.

The determinacy of algorithms

An algorithm has the comforting property that its performance does not depend on the performer, as long as the knowledge conditions (a)–(c) mentioned earlier are satisfied by the performer. For the same input to an algorithm the same output will obtain regardless of who (or what) is executing the algorithm. Moreover, an algorithm will always produce the same result regardless of when it is executed. Collectively, these two attributes imply that algorithms are *determinate*.

Which is why cookbook recipes are usually not algorithms: oftentimes they include steps that are ambiguous, thus undermining the definiteness criterion. For example, they may include instructions to add ingredients that have been 'mashed lightly' or 'finely grated' or an injunction to 'cook slowly'. These instructions are too ambiguous to satisfy the conditions of algorithm-hood. Rather, it is left to the cook's intuition, experience, and judgement to interpret such instructions. This is why the same dish prepared from the same recipe by two different cooks may differ in taste; or why the same recipe followed by the same person on two different occasions may differ in taste. Recipes violate the principle of determinacy.

Algorithms are abstract artefacts

An algorithm is undoubtedly an artefact; it is designed or invented by human beings in response to goals or needs. And insofar as they process symbol structures (as in the cases of the GCD and multiplication algorithms) they are computational. (Not all algorithms process symbol structures: the litmus test takes physical entities—a test tube of liquid, a litmus strip—as inputs and produces a physical state—the colour of the litmus strip—as output. The litmus test is a manual algorithm that operates upon physico-chemical entities, not symbol structures; we should not, then, think of it as a computational artefact.)

But algorithms, whether computational or not, themselves have no physical existence. One can neither touch nor hold them, feel them, taste them, or hear them. They obey neither the laws of physics and chemistry nor the laws of engineering sciences. They are *abstract artefacts*. They are made of symbol structures which, like all symbol structures, represent other things in the world which themselves may be physical (litmus paper, chemicals, buttons on a pocket calculator, etc.) or abstract (integers, operations on integers, relations such as equality, etc.).

An algorithm is a tool. And as in the case of most tools, the less the user needs to know its theoretical underpinnings the more effective an algorithm is for the user.

This raises the following point: as an artefact an algorithm is Janus-faced. (Janus was the Roman god of gates who looked simultaneously in two opposite directions.) Its *design* or *invention* generally demands creativity, but its *use* is a purely mechanical act demanding little creative thought. Executing an algorithm is, so to speak, a form of mindless thinking.

Algorithms are *procedural knowledge*

As artefacts, algorithms are tools users deploy to solve problems. Once created and made public they belong to the world. This is what makes algorithms objective (as well as determinate) artefacts. But algorithms are also embodiments of knowledge. And being objective artefacts they are embodiments of what philosopher of science Karl Popper called 'objective knowledge'. (It may sound paradoxical to say that the user of an algorithm is both a 'mindless thinker' yet a 'knowing subject'; but even thinking mindlessly is still thinking, and thinking entails drawing upon knowledge of some sort.) But what *kind* of knowledge does the algorithm represent?

In the natural sciences we learn definitions, facts, theories, laws, etc. Here are some examples from elementary physics and chemistry.

(i) The velocity of light in a vacuum is 186,000 miles per second.
(ii) Acceleration is the rate of change of velocity.
(iii) The atomic weight of hydrogen is 1.
(iv) There are four states of matter, solid, liquid, gas, and plasma.
(v) When the chemical elements are arranged in order of atomic numbers there is a periodic (recurring) pattern of the properties of the elements.
(vi) Combustion requires the presence of oxygen.

In each case something is *declared* to be the case: that combustion requires the presence of oxygen; that the atomic weight of hydrogen is 1; that acceleration is the rate of change of velocity; and so on. The (approximation to) truth of these statements is either by way of definition (ii); calculation (i); experimentation or observation (iv, vi); or reasoning (v). Strictly speaking, taken in isolation, they are items of information which when assimilated become part of a person's knowledge (see Chapter 1). This kind of knowledge is called *declarative knowledge* or, more colloquially, 'know-that' knowledge.

Mathematics also has declarative knowledge, in the form of definitions, axioms, or theorems. For example, a fundamental axiom of arithmetic, due to the Italian mathematician Giuseppe Peano is the principle of mathematical induction:

> Any property belonging to zero, and also to the immediate successor of every number that has that property, belongs to all numbers.

Pythagoras's theorem, in contrast, is a piece of declarative knowledge in plane geometry by way of reasoning (by proof):

> In a right-angled triangle with sides a, b forming the right angle and c the hypotenuse, the relationship $c^2 = a^2 + b^2$ is true.

Here is an example of declarative mathematical knowledge by definition:

The *factorial* of a non-negative integer n is:

$$\text{factorial}(n) = 1 \text{ for } n = 0 \text{ or } n = 1$$
$$= n(n-1)(n-2)\ldots 3.2.1 \text{ for } n > 1$$

In contrast, an algorithm is not declarative; rather it constitutes a procedure, describing how to do something. It prescribes action of some sort. Accordingly, an algorithm is an instance of *procedural knowledge* or, colloquially, 'know-how'.

For a computer scientist it is not enough to know that the factorial of a number is defined as such and such. She wants to know *how to compute* the factorial of a number. She wants an algorithm, in other words. For example:

> FACTORIAL
> **Input:** $n \geq 0$;
> **Temp variable:** *fact*;
> **Step 1:** *fact* ← 1;
> **Step 2: if** $n \neq 0$ **and** $n \neq 1$ **then**
>   **repeat**
> **Step 3:** *fact* ← *fact* * n;
> **Step 4:** $n = n - 1$;
> **Step 5: until** $n = 1$;
> **Step 6:** output *fact*;
> **Step 7:** halt

The notation **repeat** *S* **until** *C* specifies an *iteration* or *loop*. The statement(s) *S* will iteratively execute until the condition *C* is true. When this happens, the loop terminates and control flows to the statement following the iteration.

Here, in step 1, *fact* is assigned the value 1. If $n = 0$ or 1, then the condition in step 2 is not satisfied, in which case control goes directly to step 6 and the value of *fact* = 1 is output and the algorithm halts in step 7. On the other hand if *n* is neither 1 nor 0 then the *loop* indicated by the **repeat...until** segment is iteratively executed, each time decrementing *n* by 1 until the loop *termination* condition $n = 1$ is satisfied. Control then goes to step 6 which when executed outputs the value of *fact* as $n(n-1)(n-2)\ldots 3.2.1$.

Notice that the same concept—'factorial'—can be presented both declaratively (as mathematicians would prefer) and procedurally (as computer scientists would desire). In fact the declarative form provides the underlying 'theory' ('what is a factorial?') for the procedural form, the algorithm ('how do we compute it?').

In summary, algorithms constitute a form of procedural but objective knowledge.

Designing algorithms

The abstractness of algorithms has a curious consequence when we consider the *design* of algorithms. This is because, in general, design is a goal-oriented (purposive) act which begins with a set of requirements *R* to be met by an artefact *A* yet to exist, and ends with a symbol structure that represents the desired artefact. In the usual case this symbol structure is *the design D(A)* of the artefact *A*. And the designer's goal is to create *D(A)* such that if *A* is implemented according to *D(A)* then *A* will satisfy *R*.

This scenario is unproblematic when the artefact A is a material one; the design of a bridge, for example, will be a representation of the structure of the bridge in the form of engineering drawings and a body of calculations and diagrams showing the forces operating on the structure. In the case of algorithms as artefacts, however, the artefact itself is a symbol structure. Thus, to speak of the design of an algorithm is to speak of a symbol structure (the design) representing another symbol structure (the algorithm). This is somewhat perplexing.

So in the case of algorithms it is more sensible and rational to think of the design and the artefact as the same. The task of designing an algorithm is that of creating a symbol structure which *is* the algorithm A such that A satisfies the requirements R.

The operative word here is 'creating'. Designing is a creative act and, as creativity researchers have shown, the creative act is a complicated blend of reason, logic, intuition, knowledge, judgement, guile, and serendipity. And yet design theorists do talk about a 'science of design' or a 'logic of design'.

Is there, then a scientific component to the design of algorithms? The answer is: 'up to a point'. There are essentially three ways in which a 'scientific approach' enters into the design of algorithms.

To begin with, a design problem does not exist in a vacuum. It is contextualized by a body of knowledge (call it a 'knowledge space') relevant to the problem and possessed by the designer. In designing a new algorithm this knowledge space becomes relevant. For instance, a similarity between the problem at hand and the problem solved by an existing algorithm (which is part of the knowledge space) may be discovered; thus the technique applied in the latter may be transferred to the present problem. This is a case of *analogical reasoning*. Or a known design strategy may seem especially appropriate to the problem at hand, so this strategy may be attempted, though with no guarantee of its success. This is a case of *heuristic reasoning*. Or there may exist a formal theory relevant to the domain to which the problem belongs; thus the theory may be brought to bear on the problem. This is a case of *theoretical reasoning*.

In other words, forms of reasoning may be brought to bear in designing an algorithm based on a body of established or well-tried or proven knowledge (both declarative and procedural). Let us call this the *knowledge factor* in algorithm design.

But, just to come up with an algorithm is not enough. There is also the obligation to convince oneself and others that the algorithm is *valid*. This entails demonstrating by systematic reasoning that the algorithm satisfies the original requirements. I will call this the *validity factor* in algorithm design.

Finally, even if it is shown that the algorithm is valid this may not be enough. There is the question about its performance: how *good* is the algorithm? Let us call this the *performance factor* in algorithm design.

These three 'factors' all entail the kinds of reasoning, logic, and rules of evidence we normally associate with science. Let us see by way of some examples how they contribute to the science of algorithm design.

The problem of translating arithmetic expressions

There is a class of computer programs called *compilers* whose job is to translate a program written in a 'high level' programming language (that is, a language that abstracts from the features of actual physical computers; for example, Fortran or C++) into a sequence of instructions that can be directly executed (interpreted) by a specific physical computer. Such a sequence of machine-specific instructions is called 'machine code'. (Programming languages are discussed in Chapter 4.)

A classical problem faced by the earliest compiler writers (in the late 1950s and 1960s) was to develop algorithms to translate *arithmetic expressions* that appear in the program into machine code. An example of such an expression is

$$(a+b)*(c-1/d)$$

Here, +, −, *, and / are the four arithmetic operators; variables a, b, c, d and the constant number 1 are called 'operands'. An expression of this form, in which the arithmetic operators appear between its two operands is called an 'infix expression'.

The knowledge space surrounding this problem (and possessed by the algorithm designer) includes the following *rules of precedence* of the arithmetic operators:

1. In the absence of parentheses, *, / have precedence over +, −.
2. *, / have the same precedence; +, − have the same precedence.
3. If operators of the same precedence appear in an expression, then left-to-right precedence applies. That is, operators are applied to operands in order of their left-to-right occurrence.
4. Expressions within parentheses have the highest precedence.

Thus, for example in the case of the expression given earlier, the order of operators will be:

a. Perform $a + b$. Call the result $t1$.
b. Perform $1/d$. Call the result $t2$.
c. Perform $c - t2$. Call the result $t3$.
d. Perform $t1 * t3$.

On the other hand if the expression was parenthesis-free:

$$a + b * c - 1/d$$

then the order of operators would be:

i. Perform $b * c$. Call the result $t1'$.
ii. Perform $1/d$. Call the result $t2'$.
iii. Perform $a + t1'$. Call the result $t3'$.
iv. Perform $t3' - t2'$.

An algorithm can be designed to produce machine code which when executed will correctly evaluate infix arithmetic expressions according to the precedence rules. (The precise nature of the algorithm will depend on the nature of the machine-dependent instructions, an idiosyncracy of the specific physical computer.) Thus the algorithm, based on the precedence rules, draws on precise rules that are part of the knowledge space relevant to the problem. Moreover, because the algorithm is based directly on the precedence rules, arguing for the algorithm's validity will be greatly facilitated. However, as the earlier examples show, parentheses make the translation of an infix expression somewhat more complicated.

There is a notation for specifying arithmetic expressions without the need for parentheses, invented by the Polish logician Jan Lukasiewiz (1878–1956) and known, consequently, as 'Polish notation'. In one form of this notation, called 'reverse Polish', the operator immediately follows its two operands in a reverse Polish expression. The following examples show the reverse Polish form for a few infix expressions.

a. For $a + b$ the reverse Polish is $a\ b\ +$.
b. For $a + b - c$ the reverse Polish is $a\ b + c\ -$.
c. For $a + b * c$ the reverse Polish is $a\ b\ c\ * +$.
d. For $(a + b) * c$ the reverse Polish is $a\ b + c\ *$.

The evaluation of a reverse Polish expression proceeds left-to-right in a straightforward fashion, thus making the translation problem easier. The rule is that the arithmetic operators encountered are applied to their preceding operands in the order of the appearance of the operators, left-to-right. For example, in the case of the infix expression

$$(a + b) * (c - 1/d)$$

the reverse Polish form is

$$ab + c1d/ - *$$

and the order of evaluation is:

i. Perform $a\ b\ +$ and call the result $t1$. So the resulting expression is $t1\ c\ 1\ d/\ -\ *$.
ii. Perform $1\ d/$ and call the result $t2$. So the resulting expression is $t1\ c\ t2\ -\ *$.
iii. Perform $c\ t2\ -$ and call the result $t3$. So the resulting expression is $t1\ t3\ *$.
iv. Perform $t1\ t3\ *$.

Of course, programmers will write arithmetic expressions in the familiar infix form. The compiler will implement an algorithm that will first translate infix expressions into reverse Polish form and then generate machine code from the reverse Polish expressions.

The problem of converting infix expressions to reverse Polish form illustrates how a sound theoretical basis and a proven design strategy can combine in designing an algorithm that is provably correct.

The design strategy is called *recursion*, and is a special case of a broader problem solving strategy known as 'divide-and-rule'. In the latter, given a problem P, if it can be partitioned into smaller subproblems *p1, p2,..., pn*, then solve *p1, p2,..., pn* independently and then combine the solutions to the subproblems to obtain a solution for *P*.

In recursion, the problem *P* is divided into a number of subproblems *that are of the same type as P but smaller*. Each subproblem is divided into still smaller subproblems of the same type and so on until the subproblems become small and simple enough to be solved directly. The solutions of the subproblems are then combined to give solutions to the 'parent' subproblems, and these combined to form solutions to *their* parents until a solution to the original problem *P* obtains.

Consider now the problem of converting algorithmically infix expressions into reverse Polish expressions. Its basis is a set of formal rules:

Let $B = \{+, -, *, /\}$ be the set of binary arithmetic operators (that is, each operator *b* in *B* has exactly two operands). Let *a* denote an operand. For an infix expression *I* denote by *I'* its reverse Polish form. Then:

(a) If *I* is a single operand *a* the reverse Polish form is *a*.

(b) If *I1 b I2* is an infix expression where *b* is an element of *B*, then the corresponding reverse Polish expression is *I1' I2' b*.

(c) If (*I*) is an infix expression its reverse Polish form is *I'*.

The recursive algorithm constructed directly from these rules is shown later as a *function*—in the mathematical sense of this term. In mathematics, a function *F applied* to an 'argument' *x*, denoted *Fx* or *F(x)*, returns the value of the function for *x*. For example, the trigonmetric function SIN applied to the argument 90 (degrees), denoted as SIN 90, returns the value 1. The square root function symbolized as $\sqrt{}$ applied to an argument, say 4 (symbolized as $\sqrt{4}$), returns the value 2.

Accordingly, the algorithm, named here RP with an infix expression I as argument is as follows.

$$\text{RP}(I)$$

Step 1:   **if** $I = a$ **then** return a
           else
Step 2:   **if** $I = I1 \, b \, I2$
                      then return RP $(I1)$ RP $(I2) \, b$
                 else
Step 3:   **if** $I = (I1)$ **then** return RP $(I1)$
Step 4:   halt

Step 3, of the general form **if** C **then** S is a special case of the **if then else** decision form: control flows to S only if condition C is true, otherwise control flows to the statement that follows the **if then**.

The function RP can thus activate *itself* recursively with 'smaller' arguments. It is easily seen that RP is a direct implementation of the conversion rules and so, is correct by construction. (Of course, not all algorithms are so self-evidently correct; their theoretical foundation may be much more complex and their correctness must then be demonstrated by careful argument or even some form of mathematical proof; or their theoretical basis may be weak or even non-existent.)

To illustrate how the algorithm works with actual arguments, consider the following examples.

(a) Suppose $I = a + b$. Then:

RP $(a + b)$ = RP (a) RP (b) +	(by step 2)
= $ab+$	(by step 1 twice)

(b) Suppose $I = (a + b) * c$. Then:

RP $((a + b) * c)$ = RP $(a + b)$ RP (c) *	(by step 2)
= RP (a) RP (b) + RP (c) *	(by step 2)
= $ab+c*$	(by step 1 thrice)

(C) Suppose $I = (a * b) + (c - 1/d)$. Then:

$RP\ ((a * b) + (c - 1/d))$
　$= RP\ (a * b)\ RP\ (c - 1/d)\ +$　　　　　(by step 2)
　$= RP\ (a)\ RP\ (b) * RP\ (c)\ RP\ (1/d) - +$　　(by step 2 thrice)
　$= RP\ (a)\ RP\ (b) * RP\ (c)\ RP\ (1)\ RP\ (d)/ - +$　(by step 2)
　$= ab*c1d/-+$　　　　　　　　　　　　(by step 1 five times)

The 'goodness' of algorithms as utilitarian artefacts

As mentioned before, it is not enough to design a correct algorithm. Like the designer of any utilitarian artefact the algorithm designer must be concerned with how *good* the algorithm is, how efficiently it does its job. Can we *measure* the goodness of an algorithm in this sense? Can we compare two rival algorithms for the same task in some quantitative fashion?

The obvious factor of goodness will be the amount of *time* the algorithm takes to execute. But an algorithm is an abstract artefact. We cannot measure it in physical time; we cannot measure time on a real clock since an algorithm *qua* algorithm does not involve any material thing. If I as a human being execute an algorithm I suppose I could measure the amount of time I take to perform the algorithm mentally (perhaps with the aid of pencil and paper). But that is only a measure of *my* performance of the algorithm on a specific set of input data. Our concern is to measure the performance of an algorithm across all its possible inputs and regardless of who is executing the algorithm.

Algorithm designers, instead, assume that each basic step of the algorithm takes the same unit of time. Think of this as 'abstract time'. And they conceive the *size* of a problem for which the algorithm is designed in terms of the number of data items that the problem is concerned with. They then adopt two measures of algorithmic 'goodness'. One has to do with the *worst case* performance of the algorithm as a function of the size n of the problem; the other measure deals with its *average* performance, again, as a function of the problem size n. Collectively, they are called *time complexity*. (An alternative measure is the *space* complexity: the amount of (abstract) memory space required to execute the algorithm.)

The average time complexity is the more realistic goodness measure, but it demands the use of probabilities and is, thus, more difficult to analyse. In this discussion we will deal only with the worst case scenario.

Consider the following problem. I have a list of n items. Each item consists of a student name and his/her email address. The list is ordered alphabetically by name. My problem is to search the list and find the email address for a particular given name.

The simplest way to do this is to start at the beginning of the list, compare each name part of each item with the given student name, proceed along the list one by one until a match is found, and then output the corresponding email address. (For simplicity, we will assume that the student's given name is somewhere in the list.) We call this the 'linear search algorithm'.

LINEAR SEARCH

Input: *student:* an array of n entries, each entry consisting of two 'fields', denoting *name* (a character string) and *email* (a character string) respectively. For the i-th entry in *student*, denote the respective fields by *student [i].name* and *student[i].email*.

Input: *given-name:* the name being 'looked up'.

Temp variable i: an integer

Step 1: $i \leftarrow 1$;
Step 2: **while** *given-name* $\neq$ *student[i].name*
Step 3:   **do** $i \leftarrow I + 1$;
Step 4: output *student[i].email*
Step 5: halt

Here, the generic notation **while** C **do** S specifies another form of iteration: while the condition C is true repeatedly execute the statement ('loop body') S. In contrast to the **repeat** S **until** C, the loop condition is tested before the loop body is entered on each iteration.

In the worst possible case the desired answer appears in the very last (n-th) entry. So, in the worst case scenario, the **while** loop will be iterated n times. In this problem n, the number of students in the list, is the critical factor: this is the problem size.

Suppose each step takes roughly the same amount of time. In the worst case, this algorithm needs $2n + 3$ time steps to find a match. Suppose n is *very* large (say 20,000). In that case, the additional factor '3' is negligible and can be ignored. The multiplicative factor '2' though doubling n is a constant factor. What dominates is n, the problem size; it is this that might vary from one student list to another. We are interested, then, in saying something about the goodness of the algorithm in terms of the amount of (abstract) time needed to perform the algorithm as a function of this n.

If an algorithm processes a problem of size n in time kn, where k is a constant, we say that the time complexity is *of order n*, denoted as **O(n)**. This called the *Big O* notation, introduced by a German mathematician P. Bachmann in 1892. This notation gives us a way of specifying the efficiency (complexity) of an algorithm as a function of the problem size. In the case of the linear search algorithm, its worst case complexity in **O(n)**. If an algorithm solves a problem in the worst case in time kn^2, its worst case time complexity is **O(n^2)**. If an algorithm takes time k$n$log$n$ its time complexity is **O($n$log$n$)**, and so on.

Clearly, then, for the same problem of size n an **O(log$n$)** algorithm will need less time than an **O(n)** algorithm, which will need less time than an **O($n$log$n$)** algorithm, and the latter will need less time than an **O(n^2)** algorithm; the latter will be better than an **O(n^3)**

algorithm. The worst algorithms are those whose time complexity is an *exponential* function of n, such as an $O(2^n)$ algorithm. The differences in the goodness of algorithms with these kinds of time complexities, were starkly illustrated by computer scientists Alfred Aho, John Hopcroft, and Jeffrey Ullman in their influential text *The Design and Analysis of Algorithms* (1974). They showed that, assuming a certain amount of physical time to perform steps of an algorithm, in 1 minute an $O(n)$ algorithm could solve problems of size $n = 6 * 10^4$; an $O(n \log n)$ algorithm for the same problem could solve problems of size $n = 4{,}893$; an $O(n^3)$ algorithm solves the same problem but only of size $n = 39$; and an exponential algorithm of $O(2^n)$ could only solve the problem of size $n = 15$.

Algorithms can thus be placed in a hierarchy based on their Big O time complexity, with an $O(k)$ algorithm (where k is a constant) highest in the hierarchy and exponential algorithms of $O(k^n)$ lowest. Their goodness drops markedly as one proceeds down the hierarchy.

Consider the student list search problem, but this time taking into account the fact that the entries in the list are alphabetically ordered by student name. In this one can do what we *approximately* do when searching a phone book or consulting a dictionary. When we look up a directory we don't start from page one and look up each name one at a time. Instead, supposing the word whose meaning we seek in a dictionary begins with a K. We flip the pages of the dictionary to one that is roughly near the Ks. If we open the dictionary to the Ms, for example, we know we have to flip back; if we open at the Hs we have to flip forward. Taking advantage of the alphabetic ordering we *reduce* the amount of search.

This approach can be followed more exactly by way of an algorithm called *binary search*. Assuming the list has $k = 2^n - 1$ entries, in each step the middle entry is identified. If the student name so identified is alphabetically 'lower' than the given name, the algorithm will ignore the entries to the left of the middle element. It will then identify the middle entry of the right half of the list and again compare. Each time, if the name is not found, it will halve the list again and continue until a match is found.

Suppose that the list has k = 15 (i.e. $2^4 - 1$) entries. And suppose these are numbered 1 through 15. Then it can easily be confirmed that the maximum paths the algorithm will travel will be one of the following:

$8 \rightarrow 4 \rightarrow 2 \rightarrow 1$
$8 \rightarrow 4 \rightarrow 2 \rightarrow 3$
$8 \rightarrow 4 \rightarrow 6 \rightarrow 5$
$8 \rightarrow 4 -\rightarrow 6 \rightarrow 7$
$8 \rightarrow 12 \rightarrow 10 \rightarrow 9$
$8 \rightarrow 12 \rightarrow 10 \rightarrow 11$
$8 \rightarrow 12 \rightarrow 14 \rightarrow 13$
$8 \rightarrow 12 \rightarrow 14 \rightarrow 15$

Here, the list entry 8 is the middle entry. So, at most only $4 = \log_2 16$ entries will be searched before a match is found. For a list of size n the worst case performance of binary search is $\mathbf{O}\,(\log n)$, an improvement over the linear search algorithm.

The aesthetics of algorithms

The *aesthetic experience*—the quest for beauty—is found not only in art, music, film, and literature but also in science, mathematics, and even technology. 'Beauty is truth, truth beauty', began the final lines of John Keats's *Ode on a Grecian Urn* (1820). The English mathematician G.H. Hardy, echoing Keats, roundly rejected the very idea of 'ugly mathematics'.

Consider why mathematicians seek different proofs for some particular theorem. Once someone has discovered a proof for a theorem why should one bother to find another, different, proof?

The answer is that mathematicians seek new proofs of theorems when the existing ones are aesthetically unappealing. They seek beauty in their mathematics.

This applies just as much to the design of algorithms. A given problem may be solved by an algorithm which is, in some way, *ugly*—that is, clumsy, or plodding. Sometimes this is manifested in the algorithm being inefficient. So computer scientists, especially those who have a training in mathematics, seek beauty in algorithms in exactly the same sense that mathematicians seek beauty in their proofs. Perhaps the most eloquent spokespersons for an aesthetics of algorithms were the computer scientists Edsger Dijkstra from the Netherlands, C.A.R. Hoare from Britain, and Donald Knuth from the United States. As Dijkstra once put it, 'Beauty is our business'.

This aesthetic desire may be satisfied by seeking algorithms being simpler, more well structured, or using a 'deep' concept.

Consider, for example, the factorial algorithm described earlier in this chapter. This iterative algorithm was based on the definition of the factorial function as:

$$\text{fact}(n) = 1 \text{ for } n = 1 \text{ or } n = 0$$
$$= n(n-1)(n-2)\ldots 3.2.1 \text{ for } n > 1$$

But there is a recursive definition of the factorial function:

$$\text{fact}(n) = 1 \text{ for } n = 1 \text{ or } n = 0$$
$$= n * \text{fact}(n-1) \text{ for } n > 1$$

The corresponding algorithm, as a function, is:

rec-fact (n)
    if $n = 0$ **or** $n = 1$
        then return 1
        else return $n *$ rec-fact $(n-1)$

Many computer scientists would find this a more aesthetically appealing algorithm because of its clean, easily understandable, austere form and the fact that it takes advantage of the more subtle recursive definition of the factorial function. Notice that the recursive and the non-recursive (iterative) algorithms are at different levels of abstraction: the recursive version might be implemented by some variant of the non-recursive version.

Intractable ('really difficult') problems

I end this chapter by shifting focus from algorithms that solve computational problems to the computational problems themselves. In an earlier section we saw that the performance of algorithms can be estimated in terms of their time (or space) complexity. For example, in the case of the student list search problem the two algorithms (linear search and binary search), though solving the same *problem* manifested two different worst case time complexities.

But consider the so-called 'travelling salesman problem': Given a set of cities and road distances between them, can a salesman beginning at his base city visit all the cities and return to his origin such that the total distance travelled is less than or equal to some particular value? As it happens, there is no known algorithm for this problem that is less than of exponential time complexity ($O(k^n)$, for some constant k and problem size n (such as the number of cities)).

A computational problem is said to be *intractable*—'really difficult'—if all the known algorithms to solve the problem are of at least exponential time complexity. Problems for which there exist algorithms of *polynomial time complexity* (e.g. $O(n^k)$) are said to be *tractable*—that is, 'practically feasible'.

The branch of computer science that deals with the in/tractability of computational problems is a formal, mathematical domain called the *theory of computational complexity*, founded in the 1960s and early 1970s predominantly by the Israeli Michael Rabin, the Canadian Stephen Cook, and the Americans Juris Hartmanis, Richard Stearns, and Richard Karp.

Complexity theorists distinguish between two *problem classes* called **P** and **NP**, respectively. The formal (that is, mathematical) definitions of these classes are formidable, related to automata theory and, in particular, certain kinds of Turing machines (see Chapter 2), and they need not detain us here. Informally, the class **P** consists of all problems solvable in polynomial time—and these are, thus tractable. Informally, the class **NP** consists of problems for which a proposed solution, which may or may not be obtained in polynomial time, can be *checked* to be true in polynomial time. For example, the travelling salesman problem does not have a (known) polynomial time algorithmic solution but *given* a solution it can be 'easily' checked in polynomial time whether the solution is correct.

But, as noted, the travelling salesman problem is intractable. Thus, the **NP** class may contain problems believed to be intractable—although **NP** also contains the **P** class of tractable problems.

The implications of these ideas are considerable. Of particular interest is a concept called **NP**-completeness. A problem π is said to be *NP-complete* if π is in **NP** and all other problems in **NP** can be *transformed* or *reduced* in polynomial time to π. This means that if π is intractable then all other problems in **NP** are intractable. Conversely, if π is tractable, then all other problems in **NP** are also tractable. Thus, all the problems in **NP** are 'equivalent' in this sense.

In 1971, Stephen Cook introduced the concept of **NP**-completeness and proved that a particular problem called the 'satisfiability problem' is **NP**-complete. (The satisfiability problem involves

Boolean (or logical) expressions—for example the expression (a or b) and c, where the terms a, b, and c are Boolean (logical) variables having only the possible (truth) values TRUE and FALSE. The problems asks: 'Is there a set of truth values for the terms of a Boolean expression such that the value of the expression is TRUE?') Cook proved that any problem in **NP** can be reduced to the satisfiability problem which is also in **NP**. Thus, if the satisfiability problem is in/tractable then so is every other problem in **NP**.

This then raised the following question: *Are there polynomial time algorithms for all* **NP** *problems?* We noted earlier that **NP** contains **P**. But what this question asks is: Is **P** *identical to* **NP**? This is the so-called **P** = **NP** *problem*, arguably the most celebrated open problem in theoretical computer science. No one has proved that **P** = **NP**, and it is widely believed that this is not the case; that is, it is widely believed (but not yet proved) that **P** ≠ **NP**. This would mean that there are problems in **NP** (such as the travelling salesman and the satisfiability problems) that are not in **P**, hence are inherently intractable; and if they are **NP**-complete then all other problems reducible to them are also intractable. There are no practically feasible algorithms for such problems.

What the theory of **NP**-completeness tells us is that many seemingly distinct problems are *connected* in a strange sort of way. One can be transformed into another; they are equivalent to one another. We grasp the significance of this idea once we realize that a huge range of computational problems applicable in business, management, industry, and technology—'real world' problems—are **NP**-complete: if only one could construct a feasible (polynomial time) algorithm for one of these problems, one could find a feasible algorithm for all the others.

So how *does* one cope with such intractable problems? One common approach is the subject of Chapter 6.

Chapter 4
The art, science, and engineering of programming

To repeat, algorithmic thinking is central to computer science. Yet, algorithms are abstract artefacts. Computer scientists can live quite contentedly (if they so desired) in the rarified world of algorithms and never venture into the 'real world', much as 'pure' mathematicians might do. But if we desire real, physical computers to carry out computations on our behalf, if we want physical computers to not only do the kinds of computations we find too tedious (though necessary) but also those that are beyond our normal cognitive capacities, then algorithmic thinking alone does not suffice. They must be *implemented* in a form that can be communicated to physical computers, interpreted, and executed by them on their own terms rather than on human terms.

This is where programming enters the computing scene. A computer program is the specification of a desired computation in a language communicable to physical computers. The act of constructing such computations is called programming and the languages for specifying programs are called programming languages.

Programs are liminal artefacts

The concept of a program is elusive, subtle, and rather strange. For one thing, as I explain shortly, the same computation can be described at several abstraction levels depending on the language in which it is expressed, thus allowing for multiple *equivalent* programs. Secondly, a program is Janus-faced: on the one hand the program is a piece of static *text*, that is, a symbol structure that has all the characteristics of an abstract artefact. On the other hand, a program is a dynamic *process*—that is, it causes things to happen within a physical computer, and such processes consume physical time and physical space; thus it has a material substrate *upon* which it acts. Moreover, it requires a material medium *for* it to work.

Thus programs have an abstract and a material face, and for this, we may call programs *liminal* artefacts. The consequences of this liminality are both huge and controversial.

First, within the computer science community, some are drawn to the abstractness of programs and they hold the view that programs are *mathematical* objects. To them, programming is a kind of mathematical activity involving axioms, definitions, theorems, and proofs. Other computer scientists insist on its material facet and hold that programs are *empirical* objects. Programming to them is an empirical engineering activity involving the classical engineering tasks of specification of requirements, design, implementation, and conducting experiments that test and evaluate the resulting artefacts.

Secondly, the analogy between programs and the mind is often drawn. If a program is a liminal artificial object, the mind is a liminal natural object. On the one hand, mental (or cognitive) processes such as remembering, thinking, perceiving, planning, language understanding and mastering, etc., can be (and have been for centuries) examined as if the mind is a purely abstract thing interacting autonomously with the 'real' world. Yet the mind has a 'seat'. Unless one is an unrepentant dualist who completely separates mind from body, one does not believe that the mind can exist outside the brain—a physical object. And so, while some philosophers and cognitive scientists study the mind *as if* it is an abstract entity, neuroscientists seek strictly physical explanations of mental phenomena in terms of brain processes.

Indeed, the scientific study of cognition has been profoundly influenced by the analogy of mind with programs. One consequence has been the development of the branch of computer science called *artificial intelligence* (AI) which attempts to create mind-like and brain-like computational artefacts. Another has been the transformation of cognitive psychology into a broader discipline called *cognitive science* at the core of which is the hypothesis that mental processes can be modelled as program-like computations.

Thus, the intellectual influence of computer science has extended well beyond the discipline itself. Much as the Darwinian theory of evolution has extended its reach beyond biology so also because of the mind–program analogy computer science's influence has gone beyond computing itself. Or rather (as we will see in a later chapter) the very *idea* of computing has extended well beyond the scope of physical computers and automatic computation. I think it is fair to say that few sciences of the artificial have had such intellectual consequences outside their own domains.

Yet another consequence of the liminality of programs is that programs and programming are inexorably entwined with *artificial languages* called programming languages. One can design algorithms using natural language perhaps augmented with some artificial notation (as is seen in the case of the algorithms presented in Chapter 3). But no one can become a programmer without mastering at least one programming language. A pithy (if only rough) formula could well be:

Algorithms + Programming Languages = Programs

This itself has several implications.

One is the development of the theory of programming languages as a branch of computer science. Inevitably, this has led to a relationship between this theory and the science of linguistics, which is concerned with the structure of natural languages.

A second implication is the never-ending quest for the 'dream' programming language which can do what is required of any language better than any of its predecessors or its contemporary competitors. This is the activity of language design. The challenge to language designers is twofold: to facilitate communication of computations to physical computers which could then execute these computations with minimal human intervention; and also to facilitate communication with *other* human beings so that they can comprehend the computation, analyse it, criticize it, and offer suggestions for improvement, just as people do with any text. This dual challenge has been the source of an abiding obsession of computer scientists with programming and other languages of computation.

Closely related to language design is a third outcome: the study and development of programs called compilers that translate programs written in a programming language into the machine code of particular physical computers. Compiler design and implementation is yet another branch of computer science.

Finally, there has been the effort to design features of physical computers that facilitate the compiler's task. This activity is called 'language-oriented computer design' and has historically been of great interest within the branch of computer science called computer architecture (see Chapter 5).

Figure 3 shows schematically the many relationships of programs and programming with these other entities and disciplines. The entities enclosed in rectangles are contributing disciplines outside computer science; the entities enclosed in ovals are disciplines within computer science.

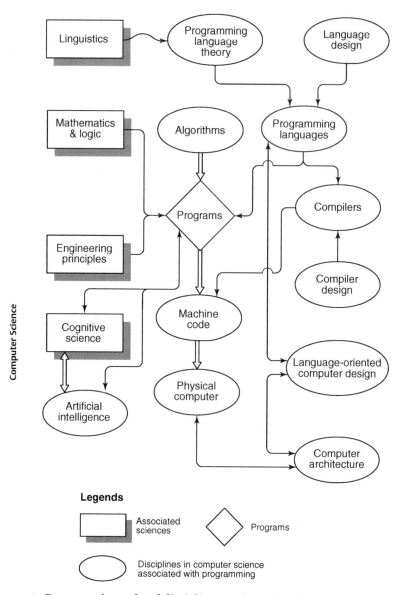

3. Programming, related disciplines, and associated sciences.

Language, thought, reality, and programming

Notice, I say 'language' in the preceding section, not 'notation'. Notation refers to symbols invented for *writing* about something. Thus, chemical or mathematical notation. Language goes beyond notation in that it affords a symbol system for *thinking* about something. Language is embroiled with thought itself.

There is a famous issue which linguists and anthropologists have debated over: is thought *defined* by language? There are those who assert that this is so, that the language we use determines the way we think about the world and what we think about, indeed, that it *defines* our conceptualization of reality itself. Its most extreme implication is that since language is a defining element of culture, thoughts, percepts or concepts are not translatable from one language culture to another. We are each trapped in our own language-defined reality. A very post-modern condition. Others take the more moderate view that language *influences,* but does not define, how we think about the world. The proposition that language defines or influences thought is called the *Sapir–Whorf hypothesis* after the anthropological linguists Edward Sapir and Benjamin Lee Whorf who developed it. (It is also called the 'principle of linguistic relativity'.)

The Sapir–Whorf hypothesis addressed natural languages. Our concern is with artificial languages, specifically those invented to express computations. No one to my knowledge has framed an analogue to the Sapir–Whorf hypothesis for the computational realm but the obsession computer scientists have shown from the earliest days of electronic computers with programming and other computational languages strongly suggest that some form of the hypothesis is tacitly accepted by the computer science community. More accurately, we may say with some confidence that the languages of computing (in particular, programming) are intimately entwined with the nature of the computing environment; and that programming languages influence programmers' mentality.

So let us consider, first, programming languages. The nature of programs will naturally emerge from this discussion.

Programming languages as abstract artefacts

It was mentioned earlier that a computation can be specified as programs at different abstraction levels that vary in 'distance' from the physical computers that can execute the programs. Correspondingly, programming languages can be conceived at different abstraction levels. A crude dichotomy separates *high-level* from *low-level* languages. The former enable programs to be written independent of all physical computers that would execute them, and the latter refers to languages designed with respect to specific families of computers or even more specifically to a particular computer.

Thus, high-level languages are 'machine-independent' and low-level ones are 'machine-dependent' with the caveat that the degree of independence or dependence may well vary. The lowest level languages are called *assembly languages* and they are so specific to particular (family of, or individual) physical computers that the assembly language programmer literally manipulates the features of the computers themselves.

There was a time in the history of computing when almost all programming was done using assembly languages. Such programs were still symbol structures (and to a very moderate extent abstract) but translators called 'assemblers' (themselves programs) would convert them into the target computers' machine code. However, because of the tedium, difficulty, amount of human time required, and error-proneness of assembly language programming the focus shifted to the invention and design of increasingly higher level, machine-independent programming languages, and the task of translating programs in such languages into executable machine code for specific computers was delegated to compilers.

In this chapter hereafter, and in the remainder of this book, unless stated explicitly, the term 'programming language' will always refer to high-level languages.

Programming languages, in contrast to natural ones, are invented or designed. They are, thus, artefacts. They entail the use of notation as symbols. As we will see, a programming language is actually a set of symbol structures and, being independent of physical computers, are abstract in exactly the same sense that algorithms are abstract. We thus have the curious situation that while programs written in such languages are liminal, the languages of programming themselves are abstract.

Language = notation + concepts & categories

Let us revisit the notation/language distinction. The world of computing is populated by hundreds of computational languages. A large majority of these are for programming but others have been created for other purposes, in particular for the design and description of physical computers at various levels of abstraction (see Chapter 5). The latter are generically called 'computer hardware description languages' (CHDLs) or 'computer design and description languages' (CDDLs).

These computational languages employ different notations— symbols—and a part of the mental effort in learning a new computational language goes into mastering the notation—that is, what the symbols symbolize. This entails mapping the notational signs onto fundamental computational *concepts* and linguistic *categories*. A particular language, then, comprises a body of concepts and categories together with the notation that represents them. Again, as a rough formula:

Concepts / Categories + Notation = Language

In computing, different signs have been deployed in different languages to symbolize the same concept. Conversely, the same sign may symbolize different concepts in different languages.

For example, the fundamental programming concept known as the *assignment* (already encountered in Chapter 3) may be denoted by such signs as '=>', '=', '←', ':=' in different languages. Thus the assignment statements:

X + 1 => X
X = X + 1
X := X + 1
X ← X + 1
X += 1

all mean the same thing: the current value of the variable X is incremented by 1 and the result is assigned to (or copied back into) X. Assignment is a computational concept, and the assignment statement is a linguistic category, and is present in most programming languages. The ways of representing it differ from one language to another depending on the taste and predilection of the language designers.

Concepts and categories in programming languages

So what *are* these concepts and categories? Computing, recall, is symbol processing; in more common parlance it is information processing. 'In the beginning is information'—or, as language designers prefer to call it, data—which is to be processed. Thus a fundamental concept embedded in all programming languages is what is called the data type. As stated in Chapter 1, a data type defines the nature of the values a data object (otherwise called 'a variable') can hold along with the operations that are permissible ('legal') on such values. Data types are either 'primitive' (or 'atomic') or 'structured' (or 'composite'), composed from more basic data types.

In the factorial example of Chapter 3, there is only the primitive data type 'non-negative integer', meaning that integers greater than or equal to 0 are admissible as values of variables of this type; and only integer arithmetic operations can be performed on variables of this type. It also means that only integer values can be assigned to variables of this type. For example, an assignment such as

$$x \leftarrow x + 1$$

is a legal statement if x is defined as of data type integer. If x has not been declared as such, if instead it is declared as (say) a character string (representing a name), then the assignment will be illegal.

A number itself, is not necessarily an integer unless it is defined as such. Thus, from a computational point of view, a telephone number is not an integer; it is a numeric character string; one cannot add or multiply two telephone numbers. So, if x is declared as a numeric character string, the earlier assignment will not be valid.

The linear search algorithm of Chapter 3 includes both primitive and structured data types. The variable i is of the primitive type integer; the variable *given* is of structured type 'character string', itself composed out of the primitive type 'character'. The list *student* is also a structured data type, sometimes called 'linear list', sometimes 'array'. The i-th element of *student* is itself a structured data type (called by different names in different programming languages, including 'record' and 'tuple') comprising here of two data types, one (*name*) of type character string, the other (*email*) also of type character string. So a variable such as *student* is a hierarchically organized *data structure*: characters composed into character strings; character strings composed into tuples or records; tuples composed into lists or arrays.

But data or information is only the beginning of a computation. Moreover the variables themselves are passive. Computation involves action and the composition of actions into processes. A programming language, thus, must not only have the facility for specifying data objects, they must also include *statements* that specify actions and processes.

In fact, we have already encountered several times the most fundamental statement *types* in the preceding chapter. One is the assignment statement: its execution invokes a process that involves both a direction and a temporal flow of information. For example, in executing the assignment statement

$$A \leftarrow B$$

where A and B are both variables, information flows from B to A. But it isn't like water flowing from one container to another. The value of B is not changed or reduced or emptied after the execution of this statement. Rather, the value of B is 'read' and 'copied' into A so that at the end, the values of A and B are equal. However, in executing the statement

$$A \leftarrow A + B$$

the value of A does change: new value of A = old value of A + value of B. The value of B remains unchanged.

The general form of the assignment statement is

$$X \leftarrow E$$

where E is an *expression* (such as the arithmetic expressions $Y + 1$, $(X - Y) * (Z/W)$). The execution of the assignment in general is a two-step process: E is first evaluated; then this value is assigned to X.

The assignment statement, then, specifies the unit of action in a computation. It is the atomic process of computations. But just as in the natural world atoms combine into molecules and molecules into larger molecules so also in the computational world. Assignments combine to form larger segments, and the latter combine to form still larger segments until complete programs obtain. There is hierarchy at work in the computational world as there is in nature.

Thus, a major task of computer scientists has been to discover the *rules of composition,* invent statement types that represent these rules, and design notations for each statement type. While the rules of composition and statement types may be quite universal, different programming languages may use different notations to represent them.

One such statement type is the *sequential* statement: two or more (simpler) statements are composed sequentially so that when executed they are executed in the order of the component statements. The notation I used in Chapter 3 to denote sequencing was the ';'. Thus, in Euclid's algorithm we find the sequential statement

$m \leftarrow n$;
$n \leftarrow r$;
goto step 1

in which the flow of control proceeds through the three statements according to the order shown.

But computations may also require making choices between one of several alternatives. The **if...then...else** statements used in Chapter 3 in several of the algorithms are instances of the *conditional* statement type in programming languages. In the general form **if** C **then** $S1$ **else** $S2$, the condition C is evaluated, and if true then control goes to $S1$, otherwise control flows to $S2$.

Sometimes, we need to return flow of control to an earlier part of the computation and repeat it. The notations **while ... do** and **repeat ... until** used in the linear search and nonrecursive factorial algorithms in Chapter 3 exemplify these instances of the *iteration* statement type.

These three statement types, the sequential, conditional, and iterative, are the building blocks for the construction of programs. Every programming language provides notation to represent these categories. In fact, in principle, any computation can be specified by a program involving a combination of just these three statement types. (This was proved in 1966 by two Italian computer theorists, Corrado Böhm and Giuseppe Jacopini.) In practice many other rules of composition and corresponding statement types have been proposed to facilitate programming (such as the *unconditional branch* exemplified by the **goto** statement used in Euclid's algorithm in Chapter 3).

Programming as art

Programming is an act of design and like all design activities, it entails judgement, intuition, aesthetic taste, and experience. It is for this that Donald Knuth entitled his celebrated and influential series of texts *The Art of Computer Programming* (1968–9). Almost a decade later Knuth elaborated on this theme in a lecture. In speaking of the art of programming, he wrote, he was alluding to programming as an *art form*. Programs should be aesthetically satisfying; they should be beautiful. The experience of writing a program should be akin to composing poetry or music. Thus, the idea of *style*, so intimately a part of the discourses of art, music, and literature, must be an element of programming aesthetics.

Recall the Dutch computer scientist Edsger Dijkstra's remark mentioned in Chapter 3: in devising algorithms, 'Beauty is our business'. The Russian computer scientist A.P. Ershov has echoed these sentiments.

Along this same theme, Knuth later proposed that programs should be *works of literature*, that one can gain pleasure in writing programs in such a way that the programs will give pleasure on being read by others. (He called this philosophy 'literate programming', though I think 'literary programming' would have been more apt as an expression of his sentiment.)

Programming as a mathematical science

But, of course, computer scientists (including Knuth) seek to discover more objective and formal foundations for programming. They want a science of programming. The Böhm–Jacopini result mentioned earlier is the sort of formal, mathematical result computer scientists yearn for. In fact, by a 'science' of programming many computer scientists mean a *mathematical* science.

The view of programming as a mathematical science has been most prominently manifested in three other ways—all having to do with the abstract face of programs or, as I noted early in this chapter, the view held by some that programs are mathematical objects.

One is the discovery of *rules of syntax* of programming languages. These are rules that determine the grammatical correctness of programs and have a huge practical bearing, since one of the first tasks of compilers (automatic translators of high level programs into machine code) is to ensure that the program it is translating is grammatically or syntactically correct. The theory of programming language syntax owes its beginnings to the linguist Noam Chomsky's work on the theory of syntax (for natural languages).

The second contribution to the science of programming is the development of *rules of semantics*—that is, principles that define the meaning of the different statement types. Its importance should be quite evident: in order to use a programming language the programmer must be quite clear about the meaning of its component statement types. So also, the compiler writer must understand unambiguously the meaning of each statement type in order to translate programs into machine code. But semantics, as the term is used in linguistics, is a thorny problem since it involves relating linguistic categories to what they refer to in the world, and the theory of programming language semantics mirrors these same difficulties. It is fair to say that the theory of semantics in programming, despite its sophisticated development, has not had the same kind of acceptance by the computer science community, nor has it been used so effectively as the theory of syntax.

The third contribution to the science of programming is closely related to the semantics issue. This contribution is founded on the conviction of such computer scientists as the Englishman C.A.R. Hoare and the Dutchman Edsger Dijkstra that computing is akin to mathematics and that the same principles that mathematicians use, such as axioms, rules of deductive logic, theorems, and proofs, are applicable to programming. This philosophy was stated quite unequivocally and defiantly by Hoare who announced, in 1985, the following manifesto:

(a) *Computers are mathematical machines.* That is, their behaviour can be mathematically defined and every detail is logically derivable from the definition.

(b) *Programs are mathematical expressions.* They describe precisely and in detail the behaviour of the computer on which they are executed.

(c) *A programming language is a mathematical theory.* It is a formal system which helps the programmer in both developing a program and proving that the program satisfies the specification of its requirements.

(d) *Programming is a mathematical activity.* Its practice requires application of the traditional methods of mathematical understanding and proof.

In the venerable axiomatic tradition in mathematics, one begins with axioms (propositions that are taken to be 'self-evidently' true about the domain of interest, such as the principle of mathematical induction mentioned in Chapter 3), and definitions of basic concepts, and *proves* progressively, new insights and propositions (collectively called theorems) from these axioms, definitions and already proved theorems, using rules of deduction. Inspired by this tradition, the third contribution to a mathematical science of programming concerns the construction of axiomatic proofs of the correctness of programs based on axioms, definitions and rules of deduction defining the semantics of the relevant programming language. The semantics is called *axiomatic semantics*, and their application is known as axiomatic proofs of correctness.

As in the axiomatic approach in mathematics (and its use in such disciplines as mathematical physics and economics), there is much formal elegance and beauty in the mathematical science of programming. However, it is only fair to point out that while a formidable body of knowledge has been developed in this realm, a goodly number of academic computer scientists and industrial practitioners remain skeptical of their practical applicability in the hurly-burly 'real world' of computing.

Programming as (software) engineering

This is because of the view held by many that programs are not 'just' beautiful abstract artefacts. Even Hoare's manifesto recognized that programs must describe the behaviour of the *computers* which execute them. The latter is the material face of programs and which confers liminality to them. To many, in fact, programs are technological products, hence programming is an engineering activity.

The word *software* appears to have entered the computing vocabulary in 1960. Yet its connotation remains uncertain. Some use 'software' and 'program' as synonyms. Some think of software to mean the special and essential set of programs (such as operating systems and other tools and 'utilities' called, collectively, 'system programs') that are built to execute atop a physical computer to create the virtual machines (or computer systems) that others can use more efficaciously (see Figure 1 in Chapter 2). Still others consider software to mean not only programs but the associated documentation that is essential to the development, operation, maintenance and modification of large programs. And there are some who would include human expertise and knowledge in this compendium.

At any rate, 'software' has the following significant connotations: it is that part of a computer system that is not itself physical; it requires the physical computer to make it operational; and there is a sense in which software is very much an *industrial* product with all that the adjective implies.

Software is, then, a computational artefact that facilitates the usage of a computer system by many (possibly millions, even billions of) users. Most times (though not always) it is a commercially produced artefact which manifests certain levels of robustness and reliability we have come to expect of industrial systems.

Perhaps in analogy with other industrial systems, a software development project is associated with a *life cycle*. And so, like many other complex, engineering projects (e.g. a new space satellite launch project) the development of a software system is regarded as an engineering project, and it is in this context that the term *software engineering* (first coined in the mid-1960s) seems particularly apposite. This has led, naturally, to the idea of the 'software engineer'. It is not a coincidence that a large portion of thinking about software engineering has originated in the industrial sector.

Various models of the software life cycle have been proposed over the past fifty years. Collectively, they all recognize that the development of a software system involves a number of stages:

(a) Analysis of the requirements the software is intended to serve.
(b) Development of precise functional, performance and cost specifications of the different components ('modules') that can be identified from requirements analysis.
(c) The design of the software system that will (hopefully) meet the specifications. This activity may itself consist of conceptual and detailed design stages.
(d) Implementing the design as an operational software system specified in a programming language and compiled for execution on the target computer system(s).
(e) Verification and validation of the implemented set of programs to ensure that they meet the specifications.
(f) Once verified and validated, the maintenance of the system and, if and whenever necessary, its modification.

These stages do not, of course, follow in a rigidly linear way. There is always the possibility of returning to an earlier stage from a later one if flaws and faults are discovered. Moreover, this software life cycle also requires an infrastructure—of tools, methodology, documentation standards, and human expertise which collectively constitute a software engineering *environment*.

One must also note that one or more of these stages will entail solid bodies of scientific theories as part of their deployment. Specification and design may involve the use of languages having their own syntax and semantics; detailed design and implementation will involve programming languages and, possibly, the use of axiomatic proof techniques. Verification and validation will invariably demand sophisticated modes of proving and experimentally testing the software. Much as the classical fields of engineering (such as structural or mechanical engineering) entail engineering sciences as components, so also software engineering.

Chapter 5
The discipline of computer architecture

The physical computer is at the bottom of the hierarchy MY_COMPUTER shown in Figure 1. In everyday parlance the physical computer is referred to as *hardware*. It is 'hard' in that it is a physical artefact that ultimately obeys the laws of nature. The physical computer is the fundamental *material* computational artefact of interest to computer scientists.

But if someone asks: 'What is the nature of the physical computer?' I may equivocate in my answer. This is because the physical computer, though a part of a larger hierarchy, is itself complex enough that it manifests its own internal hierarchy. Thus it can be designed and described at multiple levels of abstraction. The relationship between these levels combine the principles of compositional, abstraction/refinement, and constructive hierarchy discussed in Chapter 1.

Perhaps the most significant aspect of this hierarchy from a computer scientist's perspective (and we owe it, in major part, to the genius of the Hungarian-American mathematician and scientific gadfly John von Neumann to first recognize it) is the distinction of the physical computer as a symbol processing computational artefact from the physical components, obeying the laws of physics, that realize this artefact. This separation is important. As a symbol processor, the computer is abstract in exactly the same sense that software is abstract; yet, like software this abstract artefact has no existence without its physical implementation.

The view of the physical computer as an abstract, symbol processing computational artefact constitutes the computer's *architecture*. (I have used the term 'architecture' before to mean the functional structure of the virtual machines shown in Figure 1. But now, 'computer architecture' has a more specific technical connotation.) The physical (digital) components that implement the architecture—the actual hardware—constitutes its *technology*. We thus have this distinction between 'computer architecture' and (digital) technology.

There is another important aspect of this distinction. A given architecture can be implemented using different technologies. Architectures are not independent of technologies in that developments in the latter influence architecture design, but there is a certain amount of autonomy or 'degrees of freedom' the designer of computer architectures enjoy. Conversely, the design of an architecture might shape the kind of technology deployed.

To draw an analogy, consider an institution such as a university. This has both its abstract and material characteristics. The organization of the university, its various administrative and academic units, their internal structure and functions, and so on, is the analogue to a computer's architecture. One can design a university (it is, after all, an artefact), describe it, discuss and analyse it, criticize it, alter its structure, just as one can any other abstract entity. But a university is implemented by way of human and physical resources. They are the analogue to a computer's (hardware) technology. Thus, while the design or evolution of a university entails a considerable degree of autonomy, its realization can only depend on the nature, availability and efficacy of its resources (the people employed, the buildings, the equipment, the physical space, the campus structure as a whole, etc.). Conversely, the design of a university's organization will influence the kinds of resources that have to be in place.

So here are the key terms for this discussion: *computer architecture* is the discipline within computer science concerned with the design, description, analysis, and study of the logical organization, behaviour, and functional elements of a physical computer; all of that constitutes the (physical) computer's architecture. The task of the computer *architect* is to design architectures that satisfy the needs of the users of the physical computer (software engineers, programmers and algorithm designers, non-technical users) on the one hand and yet are economically and technologically viable.

Computer architectures are thus liminal artefacts. The computer architect must navigate delicately between the Scylla of the computer's functional and performance requirements and the Charybdis of technological feasibility.

Solipsistic and sociable computers

The globalization of everything owes much to the computer. If 'no man is an island entire of itself', then nor in the 21st century is the computer. But once upon a time, and for many years, computers did indeed exist as islands of their own. A computational artefact of the kind depicted as TEXT in Figure 1 would go about its tasks as if the world beyond did not exist. Its only interaction with the environment was by way of input data and commands and its output results. Other than that, for all practical purposes, the physical computer, along with its dedicated system and application programs and other tools (such as programming languages), lived in splendid, solipsistic isolation.

But, as just noted, few computers are solipsistic nowadays. The advent of the Internet, the institution of emails, the World Wide Web, and the various forms of social media have put paid to computational solipsism. Even the most reclusive user's laptop or smart phone is a sociable computer as soon as that user goes online to purchase a book or check out the weather or seek directions to go to some place. His computer is sociable in that it interacts and communicates with innumerable other computers (though blissfully unknown to him) physically located all over the planet through the network that is the Internet. Indeed, every emailer, every seeker of information, every watcher of an online video is not just using the Internet: her computer is part *of* the Internet. Or rather, the Internet is one global interactive community of sociable computational artefacts and human agents.

But there are more modest networks of which a computer can be a part. Machines within an organization (such as a university or a company) are connected to one another through what are called 'local area networks'. And a constellation of computers distributed over a region may collaborate, each performing its own computational task but exchanging information as and when needed. Such systems are traditionally called *multicomputing* or *distributed computing* systems.

The management of multicomputing or distributed computing or Internet computing is effected by a combination of network principles called 'protocols' and extremely sophisticated software systems. When we consider the discipline of computer architecture, however, it is the individual computer, whether solipsistic or sociable, that commands our attention. It is this I discuss in the remainder of this chapter.

Outer and inner architectures

The word 'architecture' in the context of computers was first used in the early 1960s by three IBM engineers, Gene Amdahl, Frederick Brooks, and Gerrit Blaauw. They used this term to mean the collection of functional attributes of a physical computer as available to the lowest level programmer of the computer (system programmers who build operating systems, compilers, and other basic utilities using assembly languages); its 'outer façade' so to speak. Since then, however, the practice of computer architecture has extended to include the internal logical, structural, and functional organization and behaviour of the physical (hardware) components of the machine. Thus, in practice, 'computer architecture' refers to the functional and logical aspects of both the outer façade and the interior of a physical computer. Yet, no agreed upon terms exist for these two aspects; here, for simplicity, I will call them 'outer' and 'inner' architectures, respectively.

The two are hierarchically related. They are two different abstractions of the physical computer, with the outer architecture an abstraction of the inner one or, conversely, the inner architecture a refinement of the outer one. Or, depending on the design strategy used, one may regard a computer's inner architecture as an implementation of the outer one.

The design of outer computer architectures is shaped by forces exerted from the computer's computational environment: since the outer architecture is the interface between the physical computer and system programmers who create the virtual machines the 'ordinary' users of the computer 'see', it is natural that the functional requirements demanded by this environment will exert an influence on the outer architectural design. For example, if the computer C is intended to support the efficient execution of programs written in a particular kind of language, say L, then the outer architecture of C may be oriented toward the features of L; thus easing the task of the language compiler in translating programs written in L into machine code for C. Or if the operating system OS that sits atop C has certain facilities, then the implementation of OS may be facilitated by appropriate features incorporated into C's outer architecture.

On the other hand, since a computer's inner architecture will be implemented by physical (hardware) components, and these

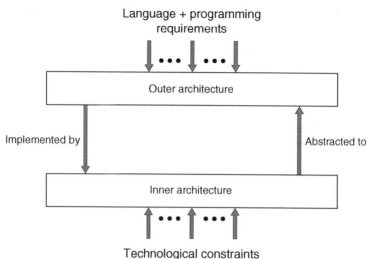

4. Computer architectures and their external constraints.

components are built using a particular kind of technology T, the design of the inner architecture will be constrained by features of T.

At the same time, the design of the outer architecture may be shaped and constrained by the nature of the inner architecture and vice versa. Thus, there is an intimate relationship between the computational environment, the outer architecture of the computer, its inner architecture, and the physical technology (Figure 4).

The outer architecture

The *sanctum sanctorum* of a computer's outer architecture is its *instruction set*, which specifies the repertoire of operations the computer can be directly commanded to perform by a programmer. Exactly what types of operations can be performed will both determine and be determined by the set of *data types* the computer directly supports or 'recognizes'. For example, if the computer is intended to efficiently support scientific and engineering computations the significant data types will be real numbers (e.g. 6.483, 4 * 10^8, − 0.000021, etc.) and integers. Thus the instruction set should include a range of arithmetic instructions.

In addition to such domain-specific instructions, there will always be a repertoire of general purpose instructions, for example, for implementing conditional (e.g. **if then else**), iterative (e.g. **while do**) and unconditional branch (e.g. **goto**) kinds of programming language constructs. Other instructions may enable a program to be organized into segments or modules responsible for different kinds of computations, with the capability of transferring control from one module to another.

An instruction is really a 'packet' that describes the operation to be performed along with references to the locations ('addresses') of its input data objects ('operands' in architectural vocabulary) and the location of where its output data will be placed. This idea of *address* implies a *memory* space. Thus, there are memory components as part of an outer architecture. Moreover, these components generally form a hierarchy:

> LONG-TERM MEMORY—known as 'backing store', 'secondary memory', or 'hard drive'.
> MEDIUM-TERM MEMORY—otherwise known as 'main memory'.
> VERY SHORT-TERM (WORKING) MEMORY—known as registers.

This is a hierarchy in terms of retentivity of information, size capacity, and speed of access. Thus, even though an outer architecture is abstract, the material aspect of the physical computer is rendered visible: space (size capacity) and time are physical, measured in physical units (bits or bytes of information, nanoseconds or picoseconds of time, etc.) not abstract. It is this combination that makes a computer architecture (outer or inner) a liminal artefact.

The long-term memory is the longest in retentivity (in its capacity to 'remember')—permanent, for all practical purposes. It is also the largest in size capacity, but slowest in access speed. Medium-term memory retains information only as long as the computer is operational. The information is lost when the computer is powered off. Its size capacity is far less than that of long-term memory but its access time is far shorter than that of long-term memory. Short-term or working memory may change its contents many times in the course of a single computation; its size capacity is several orders of magnitude lower than that of medium-term memory, but its access time is much less than that of long-term memory.

The reason for a memory hierarchy is to maintain a judicious balance between retentivity, and space and time demands for computations. There will be also instructions in the instruction set to effect transfers of programs and data between these memory components.

The other features of outer architecture are built around the instruction set and its set of data types. For example, instructions must have ways of identifying the locations ('addresses') in memory of the operands and of instructions. The different ways of identifying memory addresses are called 'addressing modes'. There will also be rules or conventions for organizing and encoding instructions of various types so that they may be efficiently held in memory; such conventions are called 'instruction formats'. Likewise, 'data formats' are conventions for organizing various data types; data objects of a particular data type are held in memory according to the relevant data format.

Finally, an important architectural parameter is the *word length*. This determines the amount of information (measured in number of bits) that can be *simultaneously* read from or written into medium-term (main) memory. The speed of executing an instruction is very much dependent on word length, as also the range of data that can be accessed per unit time.

Here are a few typical examples of computer instructions (or *machine instructions*, a term I have already used before) written in symbolic (assembly language) notation, along with their semantics (that is the actions these instructions cause to happen).

Instruction	Meaning (Action)
1. LOAD R2, (R1, D)	R2 ← main-memory [R1 + D]
2. ADD R2, 1	R2 ← R2 + 1
3. JUMP R1, D	**goto** main-memory [R1 + D]

Legend:
R1, R2: registers
main-memory: medium-term memory
1: the integer constant '1'
D: an integer number

'R1 + D' in (1) adds the number 'D' to the *contents* of register R1 and this determines the main-memory address of an operand. 'R1 + D' in (3) computes an address likewise but this address is interpreted as that of an instruction in main-memory to which control is transferred.

The inner architecture

A physical computer is ultimately a complex of circuits, wires, and other physical components. In principle, the outer architecture can be explained as the outcome of the structure and behaviour of these physical components. However, the conceptual distance between an abstract artefact like an outer architecture and the physical circuits is so large that it is no more meaningful to attempt such an explanation than it is to explain or describe a whole living organism (except perhaps bacteria and viruses) in terms of its cell biology. Cell biology does not suffice to explain, say, the structure and functioning of the cardio-vascular system. Entities above the cell level (e.g. tissues and organs) need to be understood before the whole system can be understood. So also, digital circuit theory does not suffice to explain the outer architecture of a computer, be it a laptop or the world's most powerful supercomputer.

This conceptual distance in the case of computers—sometimes called 'semantic gap'—is bridged in a hierarchic fashion. The implementation of an outer architecture is explained in terms of the inner architecture and its components. If the inner architecture is itself complex and there is still a conceptual distance from the circuit level, then the inner architecture is described and explained in terms of a still lower level of abstraction called *microarchitecture*. The latter in turn may be refined to what is called the 'logic level', and this may be sufficiently close to the circuit level that it can be implemented in terms of the latter components. Broadly speaking, a physical computer will admit of the following levels of description/abstraction:

Level 4: Outer architecture
Level 3: Inner architecture
Level 2: Microarchitecture
Level 1: Logic level
Level 0: Circuit level

Computer architects *generally* concern themselves with the outer and inner architectures, and a refinement of the inner architecture which is shown earlier as microarchitecture (this refinement is explained later). They are interested not only in the features constituting these architectural levels but also the relationship between them.

The principal components of a computer's inner architecture are shown in Figure 5. It consists of the following. First, a *memory system* which includes the memory hierarchy visible in the outer architecture but includes other components that are only visible in the inner architectural level. This system includes controllers responsible for the management of information (symbol

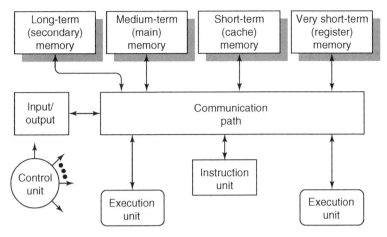

5. Portrait of a computer's inner architecture.

structures) that pass between the memories in the hierarchy, and between the system and the rest of the computer. Second, one or more *instruction interpretation units* which prepare instructions for execution and control their execution. Third, one or more *execution units* responsible for the actual execution of the various classes of instructions demanded in a computation. (Collectively, the instruction interpretation system(s) and execution unit(s) are called the computer's *processor*.) Fourth, a *communication network* that serve to transfer symbol structures between the other functional components. Fifth, an *input/output system* responsible for receiving symbol structures from, and sending symbol structures to, the physical computer's environment. And finally, the *control unit* which is responsible for issuing signals that control the activities of the other components.

The execution units are rather like the organs of a living body. They can be highly specialized for operations of specific sorts on specific data types, or more general purpose units capable of performing comprehensive sets of operations. For example, one execution unit may be dedicated to performing only integer arithmetic operations while another is specialized for arithmetic operations on real numbers; a third only manipulates bit strings in various ways; another does operations on character strings, and so on.

Internally, a processor will have its own dedicated, extremely short-term or *transient*, memory elements (of shorter term retentivity than the registers visible in the outer architecture, and sometimes called 'buffer registers') to which information must be brought from other memories before instructions can be actually processed by the instruction interpretation or processing units. Such buffer registers form the 'lowest' level in the memory hierarchy visible in the inner architecture.

There is yet another component of the memory hierarchy visible in the inner architecture but (usually) abstracted away in the outer architecture. This is a memory element called *cache* memory that lies between the medium-term (main) memory and the very short-term (register) memory. In Figure 5, this is shown as 'short-term memory'. Its capacity and speed of access lie between the two. The basic idea of a cache is that since instructions within a program module execute (usually) in sequence, a chunk of instructions can be placed in the cache so that instructions can be accessed more rapidly than if main memory is accessed. Likewise, the nature of program behaviour is such that data is also often accessed from sequential addresses in main memory so data chunks may also be placed in a cache. Only when the relevant instruction or data object is not found in the cache, will main memory be accessed, and this will cause a chunk of information in the cache to be replaced by the new chunk in which the relevant information is located so that future references to instructions and data will be available in the cache.

'The computer-within-the computer'

So how can we connect the outer to the inner architecture? How do they actually relate? To understand this we need to understand the function of the *control unit* (which, in Figure 5 stands in splendid black box-like isolation).

The control unit is, metaphorically, the computer's brain, a kind of homunculus, and is sometimes described as a 'computer-within-the-computer'. It is the organ which manages, controls, and sequences all the activities of the other systems, and the movement of symbol structures between them. It does so by issuing *control signals* (symbol structures that are categorically distinct from instructions and data) to other parts of the machine as and when required. It is the puppeteer that pulls the strings to activate the other puppet-like systems.

In particular, the control unit issues signals to the processor (the combination of the instruction interpretation and execution units) to cause a repetitive algorithm usually called the *instruction cycle* (ICycle for short) to be executed by the processor. It is the ICycle which ties the outer to the inner architecture. The general form of this is as follows:

ICYCLE:
Input: *main-memory*: medium-term memory; *registers*: short-term memory;

Internal: *pc*: transient buffer; *ir*: transient buffer; *or*: transient buffer

{*pc*, short for 'program counter', holds the address of the next instruction to be executed; *ir*, 'instruction register', holds the current instruction to be executed; *or* will hold the values of the operands of an instruction}.

FETCH INSTRUCTION: Using value of pc transfer instruction from main-memory to *ir* (*ir* ← main-memory [*pc*])

DECODE the operation part of instruction in *ir*;

CALCULATE OPERAND ADDRESSES: decode the address modes of operands in instruction in *ir* and determine the effective addresses of operands and result locations in main-memory or registers.

FETCH OPERANDS from memory system into *or*.

EXECUTE the operation specified in the instruction in *ir* using the operand values in *or* as inputs.

STORE result of the operation in the destination location for result specified in *ir*.

UPDATE PC: if the operation performed in EXECUTE is not a **goto** type operation then $pc \leftarrow pc + 1$. Otherwise do nothing: EXECUTE will have placed the address of the target **goto** instruction in *ir* into *pc*.

The ICycle is controlled by the control unit but it is the instruction interpretation unit that performs the FETCH INSTRUCTION through FETCH OPERANDS steps of the ICycle, and then the STORE and UPDATE steps, and an execution unit will perform the EXECUTE step. As a specific example consider the LOAD instruction described earlier:

LOAD R2, (R1, D)

Notice that the semantics of this instruction *at the outer architectural* level is simply

R2 ← Main-memory [R1 + D]

At the *inner architectural* level, its execution entails the performance of the ICycle. The instruction is FETCHed into ir, it is DECODEd, the operand addresses are CALCULATEd, the operands are FETCHed, the instruction is EXECUTEd, and the result STOREd into register R2. All these steps of the ICycle are abstracted out in the outer architecture as unnecessary detail as far as the users of the outer architecture (the system programmers) are concerned.

Microprogramming

To repeat, the ICycle is an algorithm whose steps are under the control of the control unit. In fact, one might grasp easily that this algorithm can be implemented *as a program executed by the control unit* with the rest of the computer (the memory system, the instruction interpretation unit, the execution units, the communication pathways, the input/output system) as part of the 'program's' *environment*. This insight, and the design of the architecture of the control unit based on this insight, was named *microprogramming* by its inventor, British computer pioneer Maurice Wilkes. And it is in the sense that the microprogrammed control unit executes a microprogram that implements the ICycle for each distinct type of instruction that led some to call the microprogrammed control unit the computer-within-the-computer. In fact, the architecture of the computer as the *microprogrammer* sees it is necessarily more refined than the inner architecture indicated in Figure 5. This microprogrammer's (or control unit implementer's) view of the computer is the 'microarchitecture' mentioned earlier.

Parallel computing

It is in the nature of those who make artefacts ('artificers')—engineers, artists, craftspeople, writers, etc.—to be never satisfied with what they have made; they desire to constantly make better artefacts (whatever the criterion of 'betterness' is). In the realm of physical computers the two dominant desiderata are space and time: to make smaller and faster machines.

One strategy for achieving these goals is by way of improving physical technology. This is a matter of solid state physics, electronics, fabrication technology, and circuit design. The extraordinary progress over the sixty or so years since the integrated circuit was first created, producing increasingly denser and increasingly smaller components and the concentration of ever increasing computing power in such components is evident to all who use laptops, tablets, and smart phones. There is a celebrated conjecture—called Moore's law, after its inventor, American engineer Gordon Moore—that the density of basic circuit components on a single chip doubles approximately every two years, which has been empirically borne out over the years.

But *given* a particular state of the art of physical technology, computer architects have evolved techniques to increase the *throughput* or *speedup* of computations, measured, for example, by such metrics as the number of instructions processed per unit time or the number of some critical operations (such as real number arithmetic operations in the case of a computer dedicated to scientific or engineering computations) per unit time. These architectural strategies fall under the rubric of *parallel processing*.

The basic idea is quite straightforward. Two processes or tasks, *T1*, *T2* are said to be executable in parallel if they occur in a *sequential task stream* (such as instructions in a sequential program) and are mutually independent. This mutual independence is achieved if they satisfy some particular conditions. The exact nature and complexity of the conditions will depend on several factors, specifically:

(a) The nature of the tasks.
(b) The structure of the task stream, e.g. whether it contains iterations (**while do** types of tasks), conditionals (**if then else**s) or **goto**s.
(c) The nature of the units that execute the tasks.

Consider, for example, the situation in which two identical processors share a memory system. We want to know under what conditions two tasks T1, T2 appearing in a sequential task stream can be initiated in parallel.

Suppose the set of input data objects to and the set of output data objects from T1 are designated as INPUT1 and OUTPUT1 respectively. Likewise, for T2, we have INPUT2 and OUTPUT2 respectively. Assume that these inputs and outputs are locations in main memory and/or registers. Then T1 and T2 can be executed in parallel (symbolized as *T1* ‖ *T2*) if *all* the following conditions are satisfied:

(a) INPUT1 and OUTPUT2 are *independent*. (That is, they have nothing in common.)
(b) INPUT2 and OUTPUT1 are independent.
(c) OUTPUT1 and OUTPUT 2 are independent.

These are known as *Bernstein's conditions* after the computer scientist A.J. Bernstein, who first formalized them. If any one of the three conditions are not met, then there is a *data dependency* relation between them, and the tasks cannot be executed in parallel. Consider, as a simple example, a segment of a program stream consisting of the following assignment statements

[1] $A \leftarrow B + C$;
[2] $D \leftarrow B * F/E$;
[3] $X \leftarrow A - D$;
[4] $W \leftarrow B - F$.

Thus:

INPUT1 = {B, C}, OUTPUT1 = {A}
INPUT2 = {B, E, F}, OUTPUT2 = {D}
INPUT3 = {A, D}, OUTPUT3 = {X}
INPUT4 = {B, F}, OUTPUT4 = {W}

Applying Bernstein's condition, it can be seen that (i) statements [1] and [2] can be executed in parallel; (ii) statements [3] and [4] can be executed in parallel; however, (iii) statements [1] and [3] have a data dependency (variable *A*); (iv) statements [2] and [3] have a data dependency constraint (variable *D*); thus these pairs cannot be executed in parallel.

So, in effect, taking these parallel and non-parallel conditions into account, and assuming that there are enough processors that can execute parallel statements simultaneously, the actual ordering of execution of the statements would be:

Statements [1] || Statement [2];
Statement [3] || Statement [4].

This sequential/parallel ordering illustrates the structure of a *parallel program* in which the tasks are individual statements described using a programming language. But consider the physical computer itself. The goal of research in parallel processing is broadly twofold: (a) Inventing algorithms or strategies that can detect parallelism between tasks and schedule or assign parallel tasks to different task execution units in a computer system. (b) Designing computers that support parallel processing.

From a computer architect's perspective the potential for parallelism exists at several levels of abstraction. Some of these *levels of parallelism* are:

(1) Task (or instruction) streams executing concurrently on independent data streams, on distinct, multiple processors but with the task streams communicating with one another (for example, by passing messages to one another or transmitting data to one another).

(2) Task (or instruction) streams executing concurrently on a single shared data stream, on multiple processors within a single computer.

(3) Multiple data streams occupying multiple memory units, accessed concurrently by a single task (instruction) stream executing on a single processor.

(4) Segments (called 'threads') of a single task stream, executing concurrently on either a single processor or multiple processors.

(5) The stages or steps of a single instruction executing concurrently within the ICycle.

(6) Parts of a microprogram executed concurrently within a computer's control unit.

All parallel processing architectures exploit variations of the possibilities just discussed, often in combination.

Consider, as an example, the abstraction level (5) given earlier. Here, the idea is that since the ICycle consists of several stages (from FETCH INSTRUCTION to UPDATE pc), the processor itself that executes the ICycle can be organized in the form of a *pipeline* consisting of these many stages. A single instruction will go through all the stages of the pipeline in sequential fashion. The 'tasks' at this level of abstraction are the steps of the ICycle through which an instruction moves. However, when an instruction occupies one of the stages, the other stages are free and they can be processing the relevant stages of other instructions in the instruction stream. Ideally, a seven-step ICycle can be executed by a seven-stage instruction processor pipeline, and all the stages of the pipeline are busy, working on seven different instructions in parallel, in an assembly line fashion. This, of course, is the ideal condition. In practice, Bernstein's conditions may be violated by instruction pairs in the instruction stream and so the pipeline may have stages that are 'empty' because of data dependency constraints between stages of instruction pairs.

Architectures that support this type of parallel processing are called *pipelined* architectures (Figure 6).

As another example, consider tasks at abstraction level (1) presented earlier. Here, multiple processors (also called 'cores')—within

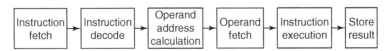

6. An instruction pipeline.

the same computer—execute instruction streams (belonging to, say, distinct program modules) in parallel. These processors may be accessing a single shared memory system or the memory system itself may be decomposed into distinct memory modules. At any rate, a sophisticated 'processor-memory interconnection network' (or 'switch') will serve as the interface between the memory systems and the processors (Figure 7). Such schemes are called *multiprocessor* architectures.

As mentioned before, the objective of parallel processing architectures is to increase the *throughput* or *speedup* of a computer system through purely architectural means. However, as the small example of the four assignment statements illustrated, there are limits to the 'parallelization' of a task stream because of data dependency constraints; thus, there are limits to the speedup that can be attained in a parallel processing environment. This limit was quantitatively formulated in the 1960s by computer designer Gene Amdahl, who stated that the potential speedup of a parallel processing computer is limited by that part of the computation that cannot be parallelized. Thus the speedup effect of increasing the number of parallel execution units levels off after a certain point. This principle is called *Amdahl's law*.

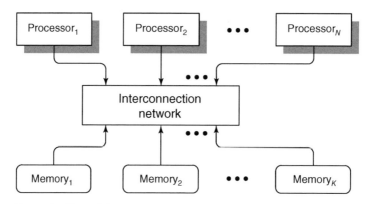

7. **Portrait of a multiprocessor.**

The science in computer architecture

The reader having reached this section of the chapter may well ask: granted that computer architectures are liminal artefacts, in what sense is the discipline a science of the artificial?

To answer this question we must recognize that the most striking aspect of the discipline is that its knowledge space consists (mainly) of a body of *heuristic* principles, and the kind of reasoning used in designing computer architectures is *heuristic reasoning*.

Heuristics—from the Greek word *hurisko*, 'to find'—are rules or propositions that offer hope or promise of solutions to certain kinds of problems (discussed in the next chapter) but *there is no guarantee of success*. To paraphrase Hungarian-American mathematician George Polya who famously recognized the role of heuristics in mathematical discovery, heuristic thinking is never definite, never final, never *certain*; rather, it is provisional, plausible, tentative.

We are often obliged to use heuristics because we may not have any other option. Heuristics are invoked in the absence of more formal, more certain, theory-based principles. The divide-and-conquer principle discussed in Chapter 3 is an example of an ubiquitous heuristic used in problem solving and decision making. It is a plausible principle which might be expected to help solve a complex problem but is not guaranteed to succeed in a particular case. Experiential knowledge is the source of many heuristics. The rule 'if it is cloudy, take an umbrella' is an example. The umbrella may well be justified but not always.

The use of heuristics brings with it the necessity of *experiment*. Since heuristics are not assured of success the only recourse is to apply them to a particular problem and see empirically if it works;

that is, conduct an experiment. Conversely, heuristic principles may themselves be derived based on prior experiments. *Heuristics and experiments go hand in hand,* an insight such pioneers in heuristic thinking as Allen Newell and Herbert Simon, and pioneers of computer design such as Maurice Wilkes fully grasped.

All this is a prelude to the following: *The discipline of computer architecture is an experimental, heuristic science of the artificial.*

Over the decades since the advent of the electronic digital computer, a body of rules, principles, precepts, propositions, and schemas have come into being concerning the design of computer architectures, almost all of which are heuristic in nature. The idea of a memory hierarchy as a design principle is an example. The principle of pipelining is another. They arise from experiential knowledge, drawing analogies, and common sense observations.

For example, experiences with prior architecture design and the difficulties faced in producing machine code using compilers have yielded heuristic principles to eliminate the difficulties. In the 1980s, the computer scientist William Wulf proposed several such heuristics based on experience with the design of compilers for certain computers. Here are some of them:

> *Regularity.* If a particular (architectural) feature is realized in a certain way in one part of the architecture then it should be realized in the same way in all parts.
>
> *Separation of concerns.* (Divide and rule.) The overall architecture should be partitionable into a number of independent features, each of which can be designed separately.
>
> *Composability.* By virtue of the two foregoing principles it should be possible to compose the separate independent features in arbitrary ways.

But experiments must follow the incorporation of heuristic principles into a design. Such experiments may entail implementing a 'prototype' or experimental machine and conducting tests on it. Or it may involve constructing a (software) simulation model of the architecture and conducting experiments on the simulated architecture.

In either case the experiments may reveal flaws in the design, in which case the outcome would be to modify the design by rejecting some principles and inserting others; and then repeat the cycle of experimentation, evaluation, and modification.

This schema is, of course, almost identical to the model of scientific problem solving advanced by philosopher of science Karl Popper:

$$P1 \rightarrow TT \rightarrow EE \rightarrow P2$$

Here, *P1* is the 'current' problem situation; *TT* is a tentative theory advanced to explain or solve the problem situation; *EE* is the process of error elimination applied to *TT* (by way of experiments and/or critical reasoning); and *P2* is the resulting new problem situation after the errors have been eliminated. In the context of computer architecture, *P1* is the design problem, specified in terms of goals and requirements the eventual computer must satisfy; *TT* is the heuristics-based design itself (which is a *theory of the computer*); *EE* is the process of experimentation and evaluation of the design, and the elimination of its flaws and limitations; and the *outcome P2* is a possibly modified set of goals and requirements constituting a new design problem.

Chapter 6
Heuristic computing

Many problems are not conducive to algorithmic solutions. A parent teaching her child to ride a bicycle cannot present to the child an algorithm he can learn and apply in the way he can learn how to multiply two numbers. A teacher of creative writing or of painting cannot offer students algorithms for writing magical realist fiction or painting abstract expressionist canvases.

This inability is in part because of one's ignorance (even that of a professor of creative writing) or lack of understanding of the exact nature of such tasks. The painter wants to capture the texture of a velvet gown, the solidity of an apple, the enigma of a smile. But what constitutes that velvetiness, that solidity, that enigmatic smile from a painterly perspective may be unknown or not known exactly enough for an algorithm to be invented for capturing them in a picture. Indeed, some would say that artistic, literary, or musical creativity can never be explicable in terms of algorithms.

Secondly, an algorithm exposes each and every step that must be followed. We can only construct an algorithm if its constitutive steps are in our conscious awareness. But so many of the actions we perform in riding a bicycle or grasping the nuances of a scene we wish to paint occur in what cognitive scientists call the 'cognitive unconscious'. There are limits to the extent such unconscious acts can be raised to the surface of consciousness.

Thirdly, even if we understand the nature of the task (reasonably) well, the task may involve multiple variables or parameters that interact with one another in nontrivial ways. Our knowledge or understanding of these interactions may be imperfect, incomplete, or hopelessly inadequate. The problem of designing a computer's outer architecture (see Chapter 5), for example, manifests this characteristic. The architect may well understand the actual parts that will go into an outer architecture (data types, operations, memory system, operand addressing modes, instruction formats, word length), but the range of variations for each such part, and the influences of these variations upon one another may be only inexactly or vaguely understood. Indeed, comprehending the full nature of these interactions may well exceed the architect's cognitive capacity.

Fourthly, even if one understands the problem well enough, and possesses knowledge about the problem domain, and can construct an algorithm to solve the problem, the amount of computational resources (time or space) needed to execute the algorithm may be simply infeasible. Algorithms of exponential time complexity (see Chapter 3) are examples.

Playing the game of chess is a case in point. The nature of the problem is very well understood. It has precise rules for legal moves, and is a game of 'perfect information' in the sense that each player can see all the pieces on the board at each point of time. The possible outcomes are precisely defined: White wins, Black wins, or they draw.

But consider the player's dilemma. Whenever it is his turn to play, his ideal objective is to choose a move that will lead to a win. In principle there is an optimum strategy (algorithm) the chess player can follow:

> The player whose turn it is to make the move considers all possible moves for himself. For each such move he then considers all

possible moves for the opponent; and for each of his opponent's possible moves, he considers again all his possible moves; and so on until an end state is reached: a win, a loss, or a draw. Then working backwards, the player determines whether the current position would force a win, and selects a move accordingly, assuming the opponent makes her moves most favourable to her.

This is called 'exhaustive search' or the 'brute force' method. In principle it will work. But of course, it is *impractical*. It has been estimated that in typical board configurations there are about thirty possible legal moves. Assume that a typical game lasts about forty moves before resignation by one of the players. Then beginning at the beginning, a player must consider thirty possible next moves; for each of these, there are thirty possible moves for the opponent, that is, 30^2 possibilities in the second move; for each of these 30^2 choices there are another thirty alternatives in the third move, that is, 30^3 possibilities. And so on until in the fortieth move the number of possibilities is 30^{40}. So at the very beginning, a player will have to consider $30 + 30^2 + 30^3 + 30^4 + \ldots + 30^{40}$ alternative moves before making an 'optimum' move. The space of alternative pathways is astronomically large.

Ultimately, if an algorithm is to be constructed to solve a problem, whatever knowledge is required about the problem for the algorithm to work *must be entirely embedded in the algorithm*. As we have noted in Chapter 3, an algorithm is a self-contained piece of procedural knowledge. To do a litmus test, perform a paper-and-pencil multiplication, evaluate the factorial of a number, generate reverse Polish expressions from infix arithmetic expressions (see Chapter 4)—all one needs to know is the algorithm itself. If one cannot incorporate any and all the necessary knowledge into the algorithm, there *is* no algorithm.

The world is full of tasks or problems manifesting the kinds of characteristics just mentioned. They include not just intellectual and creative work—scientific research, invention, designing, creative writing, mathematical work, literary analysis, historical research—but also the kinds of tasks professional practitioners—doctors, architects, engineers, industrial designers, planners, teachers, craftspeople—do. Even ordinary, humdrum activities—driving through a busy thoroughfare, making a decision about a job offer, planning a holiday trip—are not conducive to algorithmic solutions, or at least to efficient algorithmic solutions.

And yet, people go about performing these tasks and solving such problems. They do not wait for algorithms, efficient or otherwise. Indeed, if we had to wait for algorithms to solve any or all our problems then we, as a species, would have long been extinct. From an evolutionary point of view, algorithms are not all there is to our ways of thinking. And so the question arises: what other *computational* means are at our disposal to perform such tasks? The answer is to resort to a mode of computing that deploys *heuristics*.

Heuristics are rules, precepts, principles, hypotheses based on common sense, experience, judgement, analogies, informed guesses, etc., that offer promise but are not guaranteed to solve problems. We encountered heuristics in the last chapter in the discussion of computer architecture. However, to speak of computer architecture as a heuristics-based science of the artificial is one thing; to deploy heuristics in automatic computation is another. It is this latter, *heuristic computing*, that we now consider.

Search and ye *may* find

Heuristic computing embodies a spirit of adventure! There is an element of uncertainty and the unknown in heuristic computing. A problem solving agent (a human being or a computer) looking for a heuristic solution to a problem is, in effect, in a kind of *terra incognita*. And just as someone in an unknown physical territory goes into exploration or search mode so also the heuristic agent: he, she, or it *searches* for a solution to the problem, in what computer scientists call a *problem space*, never quite sure that a solution will obtain. Thus one kind of heuristic computing is also called *heuristic search*.

Consider, for example, the following scenario. You are entering a very large parking area attached to an auditorium where you wish to watch an event. The problem is to find a parking space. Cars are already parked there but you obviously have no knowledge of the distribution or location of empty spots. So what does one do?

In this case, the parking area is, literally, the problem space. And all you can do is, literally, search for an empty spot. But rather than searching aimlessly or randomly, you may decide to adopt a 'first fit' policy: pull into the first available empty spot you encounter. Or you may adopt a 'best fit' policy: find an empty spot that is the nearest to the auditorium.

These are heuristics that help *direct* the search through the problem space. There are tradeoffs, of course: the first fit strategy may reduce the search time but you may have to walk a long way to reach the auditorium; the best fit may demand a much longer search time, but if successful, the walk time may be relatively short. But, of course, *neither heuristic guarantees success*: neither is algorithmic in this sense. In both cases there may be no empty slot found, in which case you may either search indefinitely or you *terminate* the search by using a separate criterion, for example, 'exit if the search time exceeds a limit'.

Many strategies, however, that deploy heuristics have all the characteristics of an algorithm (as we discussed in Chapter 3)—with one notable difference: they give only 'nearly right' answers for a problem, or they may only give correct answers to some instances of the problem. Computer scientists, thus, refer to some kinds of heuristic problem solving techniques as *heuristic* or *approximate* algorithms, in which case we may need to distinguish them from what we may call *exact* algorithms. The term 'heuristic computing' encompasses both heuristic search and heuristic algorithms. An instance of the latter is presented shortly.

A meta-heuristic called 'satisficing'

Typically, in an optimization problem, the objective is to find the best possible solution to the problem. Many optimization problems have exact algorithmic solutions. Unfortunately, these algorithms are very often of exponential time complexity and so, impractical, even infeasible, to use for large instances of the problem. The chess problem considered earlier is an example. So what does a problem solver do, if optimal algorithms are computationally infeasible?

Instead of stubbornly pursuing the goal of optimality, the agent may aspire to achieve more feasible or 'reasonable' goals that are less than optimal but are 'acceptably good'. If a solution is obtained that meets this aspiration level then the problem solver is satisfied. Herbert Simon coined a term for this kind of mentality: *satisficing*. To satisfice is a more modest ambition than to optimize; it is to choose the feasible good over the infeasible best.

The satisficing principle is a very high level, very general, heuristic which can serve as a springboard for identifying more domain-specific heuristics. We may well call it a 'meta-heuristic'.

A chess-related satisficing heuristic. Consider the chess player's dilemma. As we have seen, optimal search is ruled out. More practical strategies are required, demanding the use of chess-related (domain-specific) heuristic principles. Among the simplest is the following.

Consider a chess board configuration (the positions of all the pieces currently on the board) **C**. Evaluate the 'promise' of **C** using some 'goodness' measure $G(\mathbf{C})$ which takes into account the general characteristics of $\mathbf{C}$ (the number and kinds of chess pieces, their relative positions, etc.).

Suppose **M1, M2, ..., Mn** are the moves that can be made by a player in configuration **C**, and suppose the resulting configurations after each such move is **M1C, M2C**, etc. Then choose a move that maximizes the goodness value of the resulting configuration. That is, choose the **Mi** whose goodness value $G(\mathbf{MiC})$ is the highest.

Notice that there is a kind of optimality attempted here. But this is a 'local' or 'short-term' optimization, looking ahead just one move. It is not a very sophisticated heuristic, but it is of a kind that the casual chess player may cultivate. But it does demand a level of deep knowledge on the player's part (whether a human being or a computer) about the relative goodness of board configurations.

Chess playing exemplifies instances of satisficing heuristic search. Consider now an instance of a satisficing heuristic algorithm.

A heuristic algorithm

Recall the discussion of parallel processing in the last chapter. A pair of tasks Ti, Tj in a task stream (at whatever level of abstraction) can be processed in parallel providing Bernstein's data independency conditions are satisfied.

Consider now a sequential stream of machine instructions generated by a compiler for a target physical computer from a high level language sequential program (see Chapter 4). If, however, the target computer can execute instructions simultaneously then the compiler has one more task to perform: to identify parallelism between instructions in the instruction stream and produce a *parallel instruction stream,* where each element of this stream consists of a set of instructions that can be executed in parallel (call this a 'parallel set').

This is, in fact, an optimization problem if the goal is to minimize the number of parallel sets in the parallel instruction stream. An optimizing algorithm would entail, like the case of the chess problem, an exhaustive search strategy and thus be computationally impractical.

In practice, more satisficing heuristics are applied. An example is what I will call here the 'First Come, First Serve' (FCFS) algorithm.

Consider the following sequential instruction stream S. (For simplicity, there are no iterations or **goto**s in this example.)

$I1: A \leftarrow B;$
$I2: C \leftarrow D + E;$
$I3: B \leftarrow E + F - 1/W;$
$I4: Z \leftarrow C + Q;$
$I5: D \leftarrow A/X;$
$I6: R \leftarrow B - Q;$
$I7: S \leftarrow D * Z.$

The FCFS algorithm is as follows.

FCFS:

> **Input**: A straight line sequential instruction stream $S:$ < I1, I2,..., In>;
>
> **Output**: A straight line parallel instruction stream P consisting of a sequence of parallel sets of instructions each and every one of which are present in S.

For each successive instruction I in S starting with I1 and ending with In, place I in the earliest possible existing parallel set subject to Bernstein's data independency conditions. If this is not possible—because of data dependency precluding placing I in any of the existing parallel sets—a new (empty) parallel set is created *after* the existing ones and I is placed there.

When this FCFS algorithm is applied to the earlier example it can be seen that the output is the parallel instruction stream *P*:

I1 || I2;
I3 || I4 || I5;
I6 || I7.

Here, '||' symbolizes parallel processing, and ';' sequential processing. This is, then, a parallel stream of three parallel sets of instructions.

FCFS is a satisficing strategy. It places each instruction in the earliest possible parallel set so that succeeding data dependent instructions can also appear as early as possible in the parallel stream. The satisficing criterion is: 'Examine each instruction on its own merit relative to its predecessors and ignore what follows'. For this particular example, FCFS produces an optimal output (a minimal sequence of parallel sets). However, there may well be other input streams for which FCFS will produce suboptimal parallel sets.

So, what is the difference between exact and heuristic algorithms? In the former case the 'goodness' is judged by evaluating its time (or space) complexity. There is no surprise or uncertainty attached to the outputs. Two or more exact algorithms for the same task (such as to solve systems of algebraic equations, sort files of data in ascending sequence, process payrolls, or compute the GCD of two integers, etc.) will not vary in their outputs; they will (or may) differ only in their performances and in their respective aesthetic appeal. In the case of heuristic algorithms there is more to the story. The algorithms may certainly be compared for their relative time complexities. (FCFS is an $O(n^2)$ algorithm for an input stream of size n.) But they may also be compared in terms of their outputs since the outputs may occasion surprise. Two parallelism detection algorithms or two chess programs employing different sets of heuristics may yield different results.

And how they differ is an empirical issue. One must implement the algorithms as executable programs, conduct experiments on various test data, examine the outputs, and ascertain their strengths and weaknesses based on the experiments. Heuristic computing, thus, entails experimentation.

Heuristics and artificial intelligence

Artificial intelligence (AI) is a branch of computer science concerned with the theory, design and implementation of computational artefacts that perform tasks we normally associate with human thinking: such artefacts can be viewed then as 'possessing' artificial intelligence. Thus it provides a bridge between computer science and psychology. And one of the earliest reflections on the possibility of artificial intelligence was due to electrical engineer Claude Shannon in the late 1940s when he considered the idea of programming a computer to play chess. In fact, ever since, computer chess has remained a significant focus in AI research. However, the most influential manifesto of what became AI (the term itself was coined by one of the pioneers of the subject, John McCarthy, in the mid-1950s) was a provocative article by Alan Turing (of Turing machine reputation) in 1950 who posed and proposed an answer to the question: 'What does it mean to claim that a computer can think?' His answer involved a kind of experiment—a 'thought experiment'—in which a human being asks questions of two invisible agents through some 'neutral' communication means (so that the interrogator cannot guess the identity of the responder from the means of response), one of the agents being a person, the other a computer. If the interrogator cannot correctly guess the identity of the computer as responder more than, say 40 per cent–50 per cent of the time then the computer may be regarded as manifesting human-like intelligence. This test came to be called the *Turing Test*, and was for some years a holy grail of AI research.

AI is a vast area and there is, in fact, more than one *paradigm* favoured by AI researchers. (I use the word 'paradigm' in philosopher of science Thomas Kuhn's sense.) Here, however, to illuminate further the scope and power of heuristic computing, I will consider only the *heuristic search paradigm* in AI.

This paradigm concerns itself with intelligent agents—natural and artificial: humans and machines. And it rests on two hypotheses articulated most explicitly by Allen Newell and Herbert Simon, the originators of the paradigm:

> *Physical Symbol System Hypothesis*: A physical symbol system has the necessary and sufficient means for general intelligent action.
>
> *Heuristic Search Hypothesis*: A physical symbol system solves problems by progressively and selectively (heuristically) searching through a problem space of symbol structures.

By 'physical symbol system', Newell and Simon meant systems that process symbol structures and yet are grounded in a physical substrate—what I have called material and liminal computational artefacts, except that they include both natural and artificial objects under their rubric.

A very general picture of a heuristic search-based problem solving agent (human or artificial) is depicted in Figure 8. A problem is solved by first creating a symbolic representation of the problem in a working memory or *problem space*. The problem representation will typically denote the *initial state*, which is where the agent starts, and the *goal state*, which represents a solution to the problem. In addition, the problem space must be capable of representing all possible states that might be reached in the effort

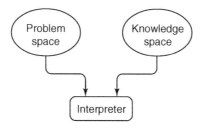

8. **General structure of a heuristic search system.**

to go from the initial to the goal state. The problem space is what mathematicians might call a 'state space'.

Transitions from one state to another are effected by appealing to the contents of the agent's *knowledge space* (the contents of a long-term memory). Elements from this knowledge space are selected and applied to a 'current state' in the problem space resulting in a new state. The organ that does this is shown in Figure 8 as the *interpreter* (on which more later). The successive applications of knowledge elements in effect result in the agent conducting a *search* for a solution through the problem space. This search process constitutes a computation. The problem is solved when, beginning with the initial state, the application of a sequence of knowledge elements results in the goal state being reached.

However, since a problem space may be arbitrarily large, the search through it is not randomly conducted. Rather, the agent deploys heuristics to control the amount of search, to prune away parts of the search space as unnecessary, and thereby converge to a solution as rapidly as possible.

Weak methods and strong methods

The heart of the heuristic search paradigm is, thus, the heuristics contained in the knowledge space. These may range from the very general—applicable to a wide range of problem domains—to the very specific—relevant to particular problem domains. The former are called *weak methods* and the latter *strong methods*. In general, when the problem domain is poorly understood weak methods are more promising; when the problem domain is known or understood in more detail strong methods are more appropriate.

One effective weak method (which we have encountered several times already) is *divide-and-rule*. Another is called *means-ends analysis:*

> Given a current problem state and a goal state, determine the difference between the two. Then reduce the difference by applying a relevant 'operator'. If, however, the necessary precondition for the operator to apply is not satisfied, reduce the difference between the current state and the precondition by recursively applying means-ends analysis to the pair, current state and precondition.

An example of a problem to which both divide-and-conquer and means-ends analysis can apply together is that of a student planning her degree program. Divide-and-conquer decomposes the problem into subproblems corresponding to each of the years of the degree program. The original goal state (to graduate in a particular subject in X number of years, say) is decomposed into 'subgoal' states for each of the X years. For each year, the student will identify the initial state for that year (the courses already taken before that year) and attempt to identify courses to be taken that eliminate the difference between the initial and goal states for that year. The search for courses is narrowed by selecting those that are mandatory. But some of these courses may require prerequisites. Thus means-ends analysis is applied to reduce the gap between the initial state and the prerequisites. And so on.

Notice that means-ends analysis is a *recursive* strategy (see Chapter 3). So what is so 'heuristic' about it? The point is that there is no guarantee that in a particular problem domain, means-ends analysis will terminate successfully. For example, given a current state and a goal state, several actions may be applicable to reduce the difference. The action chosen may determine the difference between success and failure.

Strong methods usually represent *expert knowledge* of the kind specialists in a problem domain possess through formal education, hands-on training, and experience. Computational systems that determine the molecular structure of chemicals or aid engineers in their design projects are typical instances. These heuristics are often represented in the knowledge space as rules (called *productions*) of the form:

>IF condition THEN action.

That is, if the current state in the problem space matches the 'condition' part of a production then the corresponding 'action' *may* be taken. As an example from the domain of digital circuit design (and implemented as part of a heuristic design automation system):

>IF   the goal of the circuit module is to convert a serial signal to a parallel one
>THEN use a shift register.

It is possible that the current state in the problem space is such that it matches the condition parts of several productions:

>IF condition1 THEN action1;
>IF condition2 THEN action2.
>   ....
>IF condition$M$ THEN action$M$.

In such a situation the choice of an action to take may have to be guided by a higher level heuristic (e.g. select the first matching production). This may turn out to be a wrong choice as realized later in the computation, in which case the system must 'backtrack' to a prior state and explore some other production.

Interpreting heuristic rules

Notice the 'interpreter' in Figure 8. Its task is to execute a cyclic algorithm analogous to the ICycle in a physical computer (Chapter 5):

> *Match*: Identify all the productions in the knowledge space the condition parts of which match the current state in the problem space. Collect these rules into a *conflict set*.
>
> *Select* a preferred rule from the conflict set according to a selection heuristic.
>
> *Execute* the action part of the preferred rule.
> *Goto Match*.

Apart from the uncertainty associated with the heuristic search paradigm, the other noticeable difference from algorithms (exact or heuristic) is (as mentioned before) that in the latter all the knowledge required to execute an algorithm is embedded in the algorithm itself. In contrast, in the heuristic search paradigm almost all the knowledge is located in the knowledge space (or long-term memory). The complexity of the heuristic search paradigm lies mostly in the richness of the knowledge space.

Chapter 7
Computational thinking

A certain mentality

Most sciences in the modern era—say, after the Second World War—are so technical, indeed esoteric, that their deeper comprehension remains largely limited to the specialists, the community of those sciences' practitioners. Think, for example, of the modern physics of fundamental particles. At best, when relevant, their implications are revealed to the larger public by way of technological consequences.

Yet there are some sciences that touch the imagination of those outside the specialists by way of the compelling nature of their central ideas. The theory of evolution is one such instance from the realm of the natural sciences. Its tentacles of influence have extended into the reaches of sociology, psychology, economics, and even computer science, fields of thought having nothing to do with genes or natural selection.

Among the sciences of the artificial, computer science manifests a similar characteristic. I am not referring to the ubiquitous and 'in your face' technological tools which have colonized the social world. I am referring, rather, to the emergence of a certain *mentality*.

This mentality, or at least its promise, was articulated passionately and eloquently by one of the pioneers of artificial intelligence, Seymour Papert, in his book *Mindstorms* (1980). His aim in this work, Papert announced, was to discuss and describe how the computer might afford human beings new ways of learning and thinking, not only as a practical, instrumental artefact but in much more fundamental, conceptual ways. Such influences would facilitate modes of thinking even when the thinkers were not in direct contact with the physical machine. For Papert, the computer held promise as a potential 'carrier of powerful ideas and of seeds of cultural change'. His book, he promised, would speak of how the computer could help humans fruitfully transgress the traditional boundaries separating objective knowledge and self-knowledge, and between the humanities and the sciences.

What Papert was articulating was a vision, perhaps utopian, that went well beyond the purely instrumental influence of computers and computing in the affairs of the world. This latter vision had existed from the very beginnings of automatic computation in the time of Charles Babbage and Ada, Countess of Lovelace in the mid-19th century. Papert's vision, rather, was the inculcation of a mentality that would guide, shape, and influence the ways in which a person would think about, perceive, and respond to, aspects of the world—one's inner world and the world outside—which prima facie have no apparent connection to computing—perhaps by way of analogy, metaphor, and imagination.

Over a quarter of a century after Papert's manifesto, computer scientist Jeanette Wing gave this mentality a name: *computational thinking*. But Wing's vision is perhaps more prosaic than was Papert's. Computational thinking, she wrote in 2008, entails approaches to such activities as problem solving, designing, and making sense of intelligent behaviour that draws on fundamental concepts of computing. Yet computational thinking cannot be an island of its own. In the realm of problem solving it would be akin to mathematical thinking; in the domain of design it would share features with the engineering mentality; and in understanding intelligent systems (including, of course, the mind) it might find common ground with scientific thinking.

Like Papert, Wing disassociated the mentality of computational thinking from the physical computer itself: one can think computationally without the presence of a computer.

But what does this mentality of computational thinking entail? We will see some examples later but before that let us follow AI researcher Paul Rosenbloom's interpretation of the notion of computational thinking in terms of two kinds of relationships: one is *interaction*, a concept introduced earlier (see Chapter 2) to mean, in Rosenbloom's phrase, 'reciprocal action, effect or influence' between two entities. However, interaction can signify unidirectional influence of one system A on another system B (notationally, Rosenbloom depicted this as '$A \rightarrow B$' or '$B \leftarrow A$') as well as bidirectional or mutual influence (notationally '$A \leftarrow \rightarrow B$'). By *implementation* Rosenbloom meant to 'put into effect' a system A at a higher abstraction level in terms of interacting processes within a system B at a lower level of abstraction (notationally, 'A/B'). A special case of implementation is *simulation*: B simulates A (A/B) when B acts to imitate or mimic the behaviour of A.

Using these two relationships, Rosenbloom explained, the simplest representation of computational thinking is when a computational artefact (C) influences the behaviour of a human being (H): C $\rightarrow$ H. Rosenbloom then goes further. Instead of just a human being H, suppose we consider a human simulating a computational artefact C: C/H. In this case we have the relationship C $\rightarrow$ C/H, meaning that computational artefacts influence human beings who simulate the behaviour of such artefacts. Or we may go still further: consider a human being H simulating mentally a computational artefact C which itself has implemented or is simulating the behaviour of some real world domain D: D/C/H. For example, suppose D is human behaviour. Then D/C means using a computer to simulate or model human behaviour. And D/C/H means a human being mentally simulating such a computer model of human behaviour. This leads to the following interpretation of computational thinking: C → D/C/H.

More nuanced interpretations are possible, but these interpretations in terms of interaction and implementation/simulation suffice to illustrate the general scope of computational thinking.

Computational thinking as mental skills

The most obvious influence computing can exercise on people is as a source of mental skills: a repertoire of analytical and problem solving tools which humans can apply in the course of their lives regardless of the presence or absence of actual computers. This was what Jeanette Wing had in mind. In particular, she took abstraction as the 'essence', the 'nuts and bolts' of computational thinking. But while (as we have seen throughout this book) abstraction is undoubtedly a core computational concept, computer science offers many more notions that one may assimilate and integrate into one's kit of thinking tools. I am thinking of heuristic methods, weak and strong; the idea of satisficing rather than optimizing as a realistic decision-making objective; of thinking in algorithmic terms and comprehending when and whether this is the appropriate pathway to problem solving; of the conditions and architecture of parallel processing as means for approaching multitasking endeavours; of approaching a problem situation from the 'top down' (beginning with the goal and the initial problem state, and refining the goal into simpler subgoals, and the latter into still simpler subgoals, etc.) or 'bottom up' (beginning with the goal and the lowest level building blocks and constructing a solution by composing building blocks into larger building blocks, and so on). But what is significant is that to acquire these tools of thought demands a certain level of mastery of the concepts of computer science. For Wing this entails introducing computational thinking as part of the educational curriculum from an early age.

But computational thinking entails more than analytical and problem solving skills. It encompasses a way of *imagining*, by way of seeing analogies and constructing metaphors. It is this *combination* of technical skills and imagination that, I think, Papert had in mind, and which provides the full richness of the mentality of computational thinking. We consider now some realms of intellectual and scientific inquiry where this mentality has proved to be effective.

Thinking computationally about the mind

Certainly, one of the most potent, albeit controversial, manifestations of this mentality is in thinking about thinking: the influence of computer science on cognitive psychology. Turning Turing's celebrated question—whether computers can think (the basis of AI)—on its head, cognitive psychologists consider the question: Is thinking a computational process?

The response to this question reaches back to the pioneering work of Allan Newell and Herbert Simon in the late 1950s, in their development of an information processing theory of human problem solving which combined such computational issues as heuristics, levels of abstraction, and symbol structures with logic. Much more recently it has led to the construction of models of *cognitive architecture*, most prominently by researchers such as psychologist John Anderson and computer scientists Allen Newell, John Laird, and Paul Rosenbloom. Anderson's series of models, called, generically, ACT, and that of Newell et al. called SOAR, were both strongly influenced by the basic principles of inner computer architecture (see Chapter 5). In these models the architecture of cognition is explored in terms of memory hierarchies holding symbol structures that represent aspects of the world, and the manipulation and processing of symbol structures by processes analogous to the instruction interpretation cycle (ICycle). These architectural models have been extensively investigated both theoretically and empirically as possible theories of the thinking mind at a certain abstraction level. Another kind of computationally influenced model of the mind begins with the principles of parallel processing and distributed computing, and envisions mind as a 'society' of distributed, communicating, and interacting cognitive modules. An influential proponent of this kind of mental modelling was AI pioneer Marvin Minsky. As for cognitive scientist and philosopher Margaret Boden, she titled her magisterial history of cognitive science *Mind as Machine* (2006): the mind *is* a computational device, by her account.

The computational brain

Representing or modelling the neuronal structure of the brain as a computational system and, conversely, computational artefacts as networks of highly abstract neuron-like entities has a history that reaches back to the pioneering work of mathematician Warren Pitts and neurophysiologist Warren McCulloch, and the irrepressible John von Neumann in the 1940s. Over the next sixty years a scientific paradigm called *connectionism* has evolved. In this approach, the mentality of computational thinking is expressed most specifically in the design of highly interconnected networks (hence the term 'connectionism') of very simple computational elements which collectively serve to model the behaviour of basic brain processes that are the building blocks in higher cognitive processes (such as detecting cues or recognizing patterns in visual processes). Connectionist architectures of the brain are at a lower abstraction level than the symbol processing cognitive architectures mentioned in the previous section.

The emergence of cognitive science

Symbol processing cognitive architectures of mind and connectionist models of the brain are two of the ways in which computational artefacts and the principles of computer science have influenced the shaping and emergence of the relatively new interdisciplinary field of *cognitive science*. I must emphasize that not all cognitive scientists—for instance the psychologist Jerome Bruner—take computation to be a central element of cognition. Nonetheless, the idea of understanding such activities as thinking, remembering, planning, problem solving, decision making, perceiving, and conceptualizing and understanding by way of constructing computational models and computation-based hypotheses is a compelling one; in particular, the view of computer science as a science of automatic symbol processing served as a powerful *catalyst* in the emergence of cognitive science itself. The core of Margaret Boden's history of cognitive science, mentioned in the previous section, is the development of automatic computing.

Understanding human creativity

The fascinating subject of creativity, ranging from the exceptional, historically original kind to the personal, everyday brand, is a vast topic that has attracted the professional attention of psychologists, psychoanalysts, philosophers, pedagogues, aestheticians, art theorists, design theorists, and intellectual historians and biographers; not to speak of the more self-reflexive creators themselves (scientists, inventors, poets and writers, musicians, artists, etc.). The range of approaches to, models and theories of, creativity is, accordingly, bewilderingly large, not least because of the many definitions of creativity.

But at least one community of creativity researchers has resorted to computational thinking as a *modus operandi*. They have proposed computational models and theories of the creative process that draw heavily on the principles of heuristic computing, representation of knowledge as complex symbol structures (called schemas), and the principles of abstraction. Here too, such is its compelling influence, computational thinking has afforded a common ground for the analysis of scientific, technological, artistic, literary, and musical creativity: a marriage of many cultures as Papert had hoped for.

For example, literary scholar Mark Turner has applied computational principles to the problem of understanding literary composition, just as philosopher of science and cognitive scientist Paul Thagard strove to explain scientific revolutions by way of computational models, and the present author, a computer scientist and creativity researcher, constructed a computational explanation for the design and invention of technological artefacts and ideas in the artificial sciences. The mentality of computational thinking has served as the glue that binds these different intellectual and creative cultures into one. In many of these computational studies of creativity, computer science has provided a precision of thought in which to express concepts pertaining to creativity which was formerly absent.

To take an example, the writer Arthur Koestler in his monumental work *The Act of Creation* (1964) postulated a process called 'bisociation' as the mechanism by which creative acts are effected. By bisociation, Koestler meant the coming together of two or more unconnected concepts and their blending, resulting in an original product. However, precisely how bisociation occurred remained unexplained. Computational thinking has afforded some creativity researchers (such as Mark Turner and this writer) explanations of certain bisociations in the precise language of computer science.

Understanding molecular information processing

In 1953, James Watson and Francis Crick famously discovered the structure of the DNA molecule. Thus was initiated the science of molecular biology. Its concerns included understanding and discovering such mechanisms as the replication of DNA, transcription of DNA to RNA, and translation of RNA into protein—fundamental biological processes. Thus the notion of molecules as carriers of information entered the biological consciousness. Theoretical biologists influenced by computational ideas began to model genetical processes in computational terms (which, incidentally, also led to the invention of algorithms based on genetical concepts). Computational thinking shaped what was called 'biological information processing' or, in contemporary jargon, *bioinformatics*.

Epilogue: is computer science a universal science?

Throughout this book the premise has been that computer science is a science of the artificial: that it is centred on symbol processing (or computational) artefacts; that it is a science of how things ought to be rather than how things are; that the goals of the artificers (algorithm designers, programmers, software engineers, computer architects, informaticists) must be taken into account in understanding the nature of this science. In all these respects the distinction from the natural sciences is clear.

However, in the last chapter we have seen that computational thinking serves as a bridge between the world of computational artefacts and the natural world, specifically, that of biological molecules, human cognition, and neuronal processes. Could it be, then, that computing not only affords a mentality but that, more insidiously, computation as a phenomenon embraces the natural and the artificial? That computer science is a *universal* science?

In recent years some computer scientists have thought precisely along these lines. Thus, Peter Denning has argued that computing should no longer be thought of as a science of the artificial, since information processes are abundantly found in nature. Denning and another computer scientist, Peter Freeman, have contended that in the past few decades the focus of (some computer scientists') attention has shifted from computational artefacts to information processes per se—including natural processes.

For Denning, Freeman, and yet another computer scientist Richard Snodgrass, computing is, thus, a *natural* science since computer scientists are as much in the business of discovering *how things are* (in the brain, in the living cell, and, even, in the realm of computational artefacts) as in elucidating how things ought to be. This point of view implies that computational artefacts are of the same ontological category as natural entities; or that there is no distinction to be made between the natural and the artificial. Snodgrass, in fact, invented a word to describe the natural science of computer science: 'Ergalics', from the Greek root 'ergon' ($\epsilon\rho\gamma\omega\nu$), meaning 'work'.

Paul Rosenbloom, in broad agreement with Snodgrass, but wishing to avoid a neologism, simply identified the computer *sciences* alongside the physical, life, and social sciences, as the 'fourth great scientific domain'.

The uniqueness of computer science as constituting a paradigm of its own has been an abiding theme of this book, and so Rosenbloom's thesis is consistent with this theme. The question is whether one should distinguish between the study of *natural information* processes and that of *artificial symbolic* processes. Here, the distinction between information and symbol seems justified. In the natural domain, entities do not represent anything but themselves. Entities such as neurons, or the nucleotides that are the building blocks of DNA, or the amino acids constituting proteins, do not represent anything but themselves. Thus, I find it problematic to refer to DNA processing as *symbol* processing, though to refer to these entities as carriers of non-referential *information* seems valid.

Ontologically, I think, a distinction has to be made between computer science as a science of the artificial and computer science as a natural science. In the former, human agency (in the form of goals and purpose, accessing knowledge, effecting action) is part of the science. In the latter case, agency is avowedly absent. The paradigms are fundamentally distinct.

Be that as it may, and regardless of any such possible ontological difference, what computer science has given us, as the preceding chapters have tried to show, is a remarkably distinctive way of perceiving, thinking about, and solving a breathtakingly broad spectrum of problems—spanning natural, social, cultural, technological, and economic realms. This is surely its most *original* scientific contribution to the modern world.

Further reading

The reader may wish to study the topics of the various chapters in more depth. The following list is a mix of some classic and historically influential (and still eminently readable) works and more contemporary texts; a mix of essays and historical works written for a broad readership and somewhat more technical articles.

Preface

S. Dasgupta (2014). *It Began with Babbage: The Genesis of Computer Science.* New York: Oxford University Press: esp. chapter 15.

Chapter 1: The 'stuff' of computing

S. Dasgupta (2014). *It Began with Babbage: The Genesis of Computer Science.* New York: Oxford University Press: chapters 1 & 2.
L. Floridi (2010). *Information: A Very Short Introduction.* Oxford: Oxford University Press.
D. Ince (2011). *Computer: A Very Short Introduction.* Oxford: Oxford University Press.
D.E. Knuth (1996). 'Algorithms, Programs and Computer Science' (originally published in 1966). *Selected Papers in Computer Science.* Stanford, CA: Center for the Study of Language and Information.
A. Newell, A.J. Perlis, & H.A. Simon (1967). 'What is Computer Science?' *Science,* 157, 1373–4.
A. Newell & H.A. Simon (1976). 'Computer Science as Empirical Inquiry: Symbols and Search', *Communications of the ACM,* 19, 113–26.

P.S. Rosenbloom (2010). *On Computing: The Fourth Great Scientific Domain*. Cambridge, MA: MIT Press.

Chapter 2: Computational artefacts

C.G. Bell, J.C. Mudge, & J.E. McNamara (1978). 'Seven Views of Computer Systems', pp. 1–26, in C.G. Bell, J.C. Mudge, & J.E. McNamara (ed.). *Computer Engineering: A DEC View of Hardware Systems Design*. Bedford, MA: Digital Press.

C.G. Bell, D.P. Siweorek, & A. Newell (1982). *Computer Structures: Principles and Examples*. New York: McGraw-Hill: esp. chapter 2.

S. Dasgupta (2014). *It Began with Babbage: The Genesis of Computer Science*. New York: Oxford University Press: esp. prologue and chapter 4.

J. Copeland (ed.) (2004). *The Essential Turing*. Oxford: Oxford University Press.

J. Copeland (2004). 'Computing', pp. 3–18, in L. Floridi (ed.). *Philosophy of Computing and Information*. Oxford: Blackwell.

E.W. Dijkstra (1968). 'The Structure of "THE" Multiprogramming System', *Communications of the ACM*, 11, 341–6.

E.W. Dijkstra (1971). 'Hierarchical Ordering of Sequential Processes', *Acta Informatica*, 1, 115–38.

D. Ince (2011). *The Computer: A Very Short Introduction*. Oxford: Oxford University Press.

H.H. Pattee (ed.) (1973). *Hierarchy Theory: The Challenge of Complex Systems*. New York: Braziller.

H.A. Simon (1996). *The Sciences of the Artificial* (3rd edn). Cambridge, MA: MIT Press.

A.S. Tanenbaum & H. Bos (2014). *Modern Operating Systems* (4th edn). Englewood Cliffs, NJ: Prentice-Hall.

Chapter 3: Algorithmic thinking

J. Copeland (2004). 'Computing', pp. 3–18, in L. Floridi (ed.). *Philosophy of Computing and Information*. Oxford: Blackwell.

E.W. Dijkstra (1965). 'Programming Considered as a Human Activity', *Proceedings of the 1965 IFIP Congress*. Amsterdam: North-Holland, pp. 213–17.

D.E. Knuth (1996). 'Algorithms', pp. 59–86, in D.E. Knuth. *Selected Papers on Computer Science*. Stanford, CA: Center for the Study of Language and Information.

D.E. Knuth (1997). *The Art of Computer Programming. Volume 1. Fundamental Algorithms* (3rd edn). Reading, MA: Addison-Wesley.

D.E. Knuth (2001). 'Aesthetics', pp. 91–138, in D.E. Knuth. *Things a Computer Scientist Rarely Talks About*. Stanford, CA: Center for the Study of Language and Information.

R. Sedgewick & K. Wayne (2011). *Algorithms* (4th edn). Reading, MA: Addison-Wesley.

Chapter 4: The art, science, and engineering of programming

F.P. Brooks, Jr (1975). *The Mythical Man-Month: Essays on Software Engineering*. Reading, MA: Addison-Wesley.

P. Freeman (1987). *Software Perspectives*. Reading, MA: Addison-Wesley.

C.A.R. Hoare (1985). *The Mathematics of Programming*. Oxford: Clarendon Press.

C.A.R. Hoare (2006). 'The Ideal of Program Correctness'. <http://www.bcs.org/upload/pdf/correctness.pdf>. Retrieved 28 May 2014.

D.E. Knuth (1992). *Literate Programming*. Stanford, CA: Center for the Study of Language and Information. See esp. 'Computer Programming as Art', pp. 1–16.

D.E. Knuth (2001). *Things a Computer Scientist Rarely Talks About*. Stanford, CA: Center for the Study of Language and Information. See esp. 'Aesthetics', pp. 91–138.

I. Sommerville (2010). *Software Engineering* (9th edn). Reading, MA: Addison-Wesley.

M.V. Wilkes (1995). *Computing Perspectives*. San Francisco: Morgan Kauffman. Esp. 'Software and the Programmer', pp. 87–92; 'From FORTRAN and ALGOL to Object-Oriented Languages', pp. 93–101.

N. Wirth (1973). *Systematic Programming: An Introduction*. Englewood Cliffs, NJ: Prentice Hall.

Chapter 5: The discipline of computer architecture

G.S. Almasi & A. Gottlieb (1989). *Highly Parallel Computing*. New York: The Benjamin Cummings Publishing Company.

C.G. Bell, J.C. Mudge, & J.E. McNamara (ed.) (1978). *Computer Engineering: A DEC View of Hardware Systems Design.* Bedford, MA: Digital Press.

C.G. Bell, D.P. Sieweorek, & A. Newell (1982). *Computer Structures: Principles and Examples.* New York: McGraw-Hill.

S. Dasgupta (2014). *It Began with Babbage: The Genesis of Computer Science.* New York: Oxford University Press.

S. Habib (ed.) (1988). *Microprogramming and Firmware Engineering Methods.* New York: Van Nostrand Reinhold.

C. Hamachar, Z. Vranesic, & S. Zaky (2011). *Computer Organization and Embedded Systems* (5th edn). New York: McGraw-Hill.

K. Hwang & F.A. Briggs (1984). *Computer Architecture and Parallel Processing.* New York: McGraw-Hill.

D.E. Ince (2011). *The Computer: A Very Short Introduction.* Oxford: Oxford University Press.

D.A. Patterson & J.L. Henessy (2011). *Computer Architecture: A Quantitative Approach* (5th edn). Burlington, MA: Morgan Kaufmann.

A.S. Tanenbaum (2011). *Structured Computer Organization* (6th edn). Englewood Cliffs, NJ: Prentice Hall.

Chapter 6: Heuristic computing

D.R. Hofstadter (1999). *Gödel, Escher, Bach: An Eternal Golden Braid* (20th anniversary edn). New York: Basic Books.

A. Newell & H.A. Simon (1972). *Human Problem Solving.* Englewood Cliffs, NJ: Prentice Hall.

A. Newell & H.A. Simon (1976). 'Computer Science as Empirical Inquiry: Symbols and Search', *Communications of the ACM*, 19, 113–26.

J. Pearl (1984). *Heuristics: Intelligent Search Strategies for Computer Problem Solving.* Reading, MA: Addison-Wesley.

G. Polya & J.H. Conway (2014). *How to Solve It: A New Aspect of Mathematical Method.* Princeton, NJ: Princeton University Press. (Originally published in 1949).

D.L. Poole & A.K. Mackworth (2010). *Artificial Intelligence: Foundations of Computational Agents.* Cambridge: Cambridge University Press.

S. Russell & P. Norvig (2014). *Artificial Intelligence: A Modern Approach* (3rd edn). New Delhi: Dorling Kinderseley/Pearson.

H.A. Simon (1995). 'Artificial Intelligence: An Empirical Science', *Artificial Intelligence* 77, 1, 95–127.

H.A. Simon (1996). *The Sciences of the Artificial* (3rd edn). Cambridge, MA: MIT Press.

Chapter 7: Computational thinking

M.A. Boden (2006). *Minds as Machines. Volume 1.* Oxford: Clarendon Press.
S. Dasgupta (1994). *Creativity in Invention and Design: Computational and Cognitive Explorations of Technological Originality.* New York: Cambridge University Press.
S. Papert (1980). *Mindstorms.* New York: Basic Books.
P.S. Rosenbloom (2013). *On Computing: The Fourth Great Scientific Domain.* Cambridge, MA: MIT Press.
J. Searle (1984). *Minds, Brains and Science.* Cambridge, MA: Harvard University Press.
J. Setubal & J. Meidinis (1997). *Introduction to Computational Molecular Biology.* Pacific Grove, CA: Brooks/Cole Publishing Company.
C.A. Stewart (ed.) (2004). '*Bioinformatics:* Transforming Biomedical Research and Medical Care' [Special Section on Bioinformatics], *Communications of the ACM*, 47/11, 30–72.
P.R. Thagard (1988). *Computational Philosophy of Science.* Cambridge, MA: MIT Press.
P.R. Thagard (1992). *Conceptual Revolutions.* Princeton, NJ: Princeton University Press.
J.M. Wing (2006). 'Computational Thinking', *Communications of the ACM*, 49/3, 33–5.
J.M. Wing (2008). 'Computational Thinking and Thinking about Computing', *Philosophical Transactions of the Royal Society*, Series A, 366, pp. 3717–25.

Epilogue: is computer science a universal science?

S. Dasgupta (2014). *It Began with Babbage: The Genesis of Computer Science.* New York: Oxford University Press.
P.J. Denning & C.H. Martell (2015). *Great Principles of Computing.* Cambridge, MA: MIT Press.
P.J. Denning (2005). 'Is Computer Science Science?' *Communications of the ACM*, 48/4, 27–31.
P.J. Denning & P.A. Freeman (2009). 'Computing's Paradigm', *Communications of the ACM*, 52/12, 28–30.

P.S. Rosenbloom (2013). *On Computing: The Fourth Great Scientific Domain*. Cambridge, MA: MIT Press.

R. Snodgrass (2010). 'Ergalics: A Natural Science of Computing'. <http://citeseerx.ist.psu.edu/viewdoc/download?doi=10.1.1.180.4704&rep=rep1&type=pdf>. Retrieved 16 Sept. 2015.

"牛津通识读本"已出书目

古典哲学的趣味	福柯	地球
人生的意义	缤纷的语言学	记忆
文学理论入门	达达和超现实主义	法律
大众经济学	佛学概论	中国文学
历史之源	维特根斯坦与哲学	托克维尔
设计,无处不在	科学哲学	休谟
生活中的心理学	印度哲学祛魅	分子
政治的历史与边界	克尔凯郭尔	法国大革命
哲学的思与惑	科学革命	民族主义
资本主义	广告	科幻作品
美国总统制	数学	罗素
海德格尔	叔本华	美国政党与选举
我们时代的伦理学	笛卡尔	美国最高法院
卡夫卡是谁	基督教神学	纪录片
考古学的过去与未来	犹太人与犹太教	大萧条与罗斯福新政
天文学简史	现代日本	领导力
社会学的意识	罗兰·巴特	无神论
康德	马基雅维里	罗马共和国
尼采	全球经济史	美国国会
亚里士多德的世界	进化	民主
西方艺术新论	性存在	英格兰文学
全球化面面观	量子理论	现代主义
简明逻辑学	牛顿新传	网络
法哲学:价值与事实	国际移民	自闭症
政治哲学与幸福根基	哈贝马斯	德里达
选择理论	医学伦理	浪漫主义
后殖民主义与世界格局	黑格尔	批判理论

德国文学	儿童心理学	电影
戏剧	时装	俄罗斯文学
腐败	现代拉丁美洲文学	古典文学
医事法	卢梭	大数据
癌症	隐私	洛克
植物	电影音乐	幸福
法语文学	抑郁症	免疫系统
微观经济学	传染病	银行学
湖泊	希腊化时代	景观设计学
拜占庭	知识	神圣罗马帝国
司法心理学	环境伦理学	大流行病
发展	美国革命	亚历山大大帝
农业	元素周期表	气候
特洛伊战争	人口学	第二次世界大战
巴比伦尼亚	社会心理学	中世纪
河流	动物	工业革命
战争与技术	项目管理	传记
品牌学	美学	公共管理
数学简史	管理学	社会语言学
物理学	卫星	物质
行为经济学	国际法	学习
计算机科学